T0332034

Project Design for Geomatics Engineers and Surveyors

Project Design for Geomatics Engineers and Surveyors, Second Edition, continues to focus on the key components and aspects of project design for geomatics and land surveying projects with the goal of helping readers navigate the priority issues when planning new projects. The second edition includes new materials on surveying and UAV, and it is thoroughly updated to keep current with the recent technology and terminology. The two new chapters capture new developments in the rapidly emerging use of remote sensing and GIS in aerial surveys, mapping, and imaging for small-to-medium-scale projects, as well as modern practices and experiences in engineering surveying.

- Provides a simple guide for geomatics engineering projects using recent and advanced technologies.
- Includes new content on spatial data collection using GIS, drones, and 3D digital modeling.
- Covers professional standards; professional and ethical responsibilities; and policy, social, and environmental issues related.
- Discusses project planning including scheduling and budgeting.
- Features practical examples with solutions and explains new methods for planning, implementing, and monitoring engineering and mining surveying projects.

Undergraduate and graduate students, professors, practicing professionals, and surveyors will find this new edition useful, as well as geospatial/geomatics engineers, civil engineers, mining engineers, GIS professionals, planners, land developers, and project managers.

Project Design for Geomatics Engineers and Surveyors

Second Edition

Clement A. Ogaja, Nashon J. Adero, and Derrick Koome

CRC Press
Taylor & Francis Group
Boca Raton London New York

CRC Press is an imprint of the
Taylor & Francis Group, an **informa** business

Second edition published 2023
by CRC Press
6000 Broken Sound Parkway NW, Suite 300, Boca Raton, FL 33487-2742

and by CRC Press
4 Park Square, Milton Park, Abingdon, Oxon, OX14 4RN

CRC Press is an imprint of Taylor & Francis Group, LLC

First edition published by CRC Press 2020

Library of Congress Cataloging-in-Publication Data
Names: Ogaja, Clement A., author. | Adero, Nashon Juma, author. | Koome, Derrick, author.
Title: Project design for geomatics engineers and surveyors / Clement Ogaja, Nashon Adero, Derrick Koome.
Other titles: Geomatics engineering
Description: Second edition. | Boca Raton : CRC Press, 2023. |
Revised edition of: Geomatics engineering : a practical guide to project design / Clement A. Ogaja. Boca Raton : CRC Press, c2011. | Includes bibliographical references and index. |
Summary: "Project Design for Geomatics Engineers and Surveyors, Second Edition", continues to focus on the key components and aspects of project design for geomatics and land surveying projects with the goal of helping readers navigate the priority issues when planning new projects. The second edition includes new materials on surveying and UAV, and it is thoroughly updated to keep current with the recent technology and terminology. The two new chapters capture new developments in the rapidly emerging use of remote sensing and GIS in aerial surveys, mapping, and imaging for small-to-medium scale projects, as well as modern practices and experiences in engineering surveying"–Provided by publisher.
Identifiers: LCCN 2022046427 (print) | LCCN 2022046428 (ebook) |
ISBN 9781032266794 (hbk) | ISBN 9781032285160 (pbk) | ISBN 9781003297147 (ebk)
Subjects: LCSH: Surveying. | Geomatics. | Project management. | Geographic information systems.

ISBN: 978-1-032-26679-4 (hbk)
ISBN: 978-1-032-28516-0 (pbk)
ISBN: 978-1-003-29714-7 (ebk)

DOI: 10.1201/9781003297147

Typeset in Times
by codeMantra

Printed and bound by CPI Group (UK) Ltd, Croydon, CR0 4YY

Contents

PART I Overview

PART II   Contemporary Issues

PART III   Planning and Design

PART IV   Proposal Development

PART V   Appendices

Preface (Second Edition)

This is a second edition of the book titled *Geomatics Engineering: A Practical Guide to Project Design*. The goal of this second edition is to build on the first edition with the theme of project design processes, professionalism, and pertinent issues, and not necessarily on the theory of topical areas of the geomatics discipline. New material, including two chapters, a section, and an appendix, has been added and existing chapters have been updated to keep them current with the updated information and terminology. All grammatical and typographical errors have been diligently corrected throughout all the chapters.

The second edition retains the objectives outlined in the first edition. The key components and aspects of project design for geomatics and land surveying projects are presented with the goal of helping the reader navigate the priority areas of attention when planning new projects. It guides readers through the project design and request for proposal process commonly used for soliciting professional geomatics and land surveying services. To better align with this objective and for a better representation of the subject matter, the book title has been changed to *Project Design for Geomatics Engineers and Surveyors, Second Edition*.

The Table of Contents is updated to add two new chapters and updates to existing chapters as well as other necessary minor changes such as the use of the acronym "GNSS" to replace "GPS", where necessary, for example in the text of Chapters 1, 2, 6, 7, 8 and 11. A glossary of terms used in geomatics and survey projects is also added in the appendix. New information has been added in Chapter 8 to discuss other sources of GIS data (Section 8.5).

The two new chapters are intended to capture new developments in, and the rapidly emerging use of, unmanned aerial systems (UASs) for aerial surveys/mapping/imagery involving small-to-medium-scale projects, as well as the modern practice and experiences from engineering and mining surveys. Mining surveys are gaining currency in a world growing more conscious of the escalating need to balance mineral extraction with environmental and social responsibility for sustainability, actionable location-based intelligence, and shared visual maps at scale, being key elements of sound multicriteria decision support.

Use of the term "GNSS" instead of "GPS" is necessary because GNSS (Global Navigation Satellite System) implies a multi-constellation global navigation satellite system which includes GPS (American GNSS), GLONASS (Russian GNSS), Galileo (European GNSS), and BeiDou (Chinese GNSS). GNSS receivers and products are now commonplace as used by surveyors and geomatics professionals.

Preface (First Edition)

Most of the courses in surveying and geomatics engineering curriculum have been designed for the students to develop a progressively increasing knowledge base and related practical skills in specific fields, such as land surveying, geodetic surveying, GIS, and photogrammetry. In design and senior design classes, students learn how to synthesize the knowledge and skills acquired in several different courses toward the planning, design, implementation, and management of comprehensive geomatics engineering projects. This requires that they understand the scope of work, correctly interpret the required standards and specifications for accuracy, and the scheduling and budgetary constraints. Students learn how to evaluate design requirements as well as economic and social considerations. A 2009 survey of books available reveals the lack of any text devoted to principles of design and professionalism in surveying and geomatics engineering.

This text, therefore, has been written to focus attention on (1) the overall project design process including scheduling and budgetary constraints; (2) standards and specifications for accuracy; (3) professionalism and ethical responsibilities; (4) policy, social, global, and environmental considerations; (5) project cost estimating process; and (6) writing of proposals in response to the request for proposal (RFP) process commonly used for soliciting professional geomatics engineering services. It is intended to introduce readers to some of the issues in solving modern geomatics engineering problems and to provide the practitioner with a frame of reference.

The nature of the book makes it a senior- or graduate-level text, and it has been written for those who already have a basic understanding of material that appears in any undergraduate book on land surveying and geomatics engineering. A complete explanation of theory, measurement, or conduct of field procedures is beyond its scope. Readers unfamiliar with such theory or procedures should consult appropriate sources of information, peers, or professionals for assistance.

The book is organized into four parts, and each chapter includes exercises to help engage in critical thinking and problem solving:

- Part I reflects, as much as possible, the natural progression of project design considerations, including how the planning, information gathering, design, scheduling, cost estimating, and proposal writing fit into the overall scheme of project design process.
- Part II presents the details of contemporary issues such as standards and specifications; professional and ethical responsibilities; and policy, social, and environmental issues that are pertinent to geomatics engineering projects.
- Part III shows the important considerations when planning or designing new projects. Although the primary goal is to demonstrate planning and design considerations for the entire field of surveying and geomatics engineering, it has been necessary to be selective and to give greater weight only to some topics.

- Part IV focuses on the proposal development process and shows how to put together a project cost estimate, including estimating quantities and developing unit and lump-sum costs.

Few books are written that include only the ideas of the author, and this book is no exception. The education and support I have received from the following institutions is almost immeasurable: the University of Nairobi (Kenya), the University of New South Wales (Australia), Geoscience Australia, and California State University (Fresno). I also acknowledge the help and support from my colleagues and students of the California State University and the support of individuals from other organizations, in particular, the assistance provided by Belle Craig and Jerry Wahl (both of BLM of the U.S. Department of Interior). Last, but not least, I am most indebted to my family (wife Julie, daughter Alicia, and son Joshua) for their never-ending support, patience, and understanding.

Acknowledgments

I would like to thank my family (wife Julie, daughter Alicia and son Joshua) for tolerating the time spent away from them on evenings, nights, and weekends while preparing this book; and above all, to Almighty God for being the source of strength and inspiration in everything I do. Much appreciation to my two co-authors for contributing two new chapters, and to anonymous reviewers for their time spent reading the manuscript and providing valuable comments and suggestions. I would like to extend my sincere thanks to all individuals and various institutions for copyright permissions to use their images and artwork in both the first edition and this edition; the institutions, teachers, and professors who provided education that shaped my life; various employers and co-workers who provided opportunities to contribute and learn; my parents for bringing me into this earth; and all who provided moral support and encouragement.

Clement A. Ogaja

Authors

Clement A. Ogaja has worked in various capacities as a professor, researcher, and geodesist in the United States, Australia, and Kenya. He earned the B.Sc. degree in surveying from the University of Nairobi in 1997 and Ph.D. in geomatics engineering from UNSW Sydney, Australia, in 2002. In addition to his three books, *Geomatics Engineering: A Practical Guide to Project Design* (CRC Press), *Applied GPS for Engineers and Project Managers* (ASCE Press), and *Introduction to GNSS Geodesy: Foundations of Precise Positioning Using Global Navigation Satellite Systems* (Springer), he is the author and co-author of several research papers published in international scientific journals and conference proceedings. He has extensive experience in education, research, and private industry having worked for Geoscience Australia, California State University, Topcon Positioning Systems, NOAA's National Geodetic Survey, and as a land surveyor with Aerophoto Systems Engineering Company in Kenya. Dr. Ogaja's primary interest is researching the applications of GPS/GNSS and Geomatics technologies to solving engineering and societal problems.

Nashon J. Adero, a lecturer at Taita Taveta University, Kenya, is a geospatial and systems modelling expert. He teaches core units of Engineering Surveying, Mine Surveying and GIS, and support courses in facility management, sustainability, communication in science and engineering, and research methodology. He has acquired extensive cross-sector experience as a tunnel surveyor, a GIS manager, and a lead consultant for government agencies and private organizations. He also trained and worked as a policy analyst at the Kenya Institute for Public Policy Research and Analysis (KIPPRA), under a postgraduate training program in economic modeling and public policy research and analysis. His doctoral research at the Technical University and Mining Academy at Freiberg focused on mining surveys and geospatial models for mine planning. He earned an M.Sc. degree in resources engineering from Karlsruhe Institute of Technology (KIT) in 2006 and B.Sc. (Hons.) degree in surveying from the University of Nairobi in 2000. He is author, co-author, and co-editor of several articles, books and book chapters, conference proceedings, policy papers, technical reports, and media articles on geospatial technologies, environmental sustainability, decision support models, and skills development. In 2021, he co-edited *The Future of Africa in the Post-COVID-19 World*, a peer-reviewed book.

Derrick Koome earned a B.Sc. degree in geospatial engineering from the University of Nairobi. He briefly worked in a busy geospatial firm in Nairobi (Nile Surveys and Geo-solutions) and by the time he left he was its Chief Surveyor. He started his own land survey practice soon after (Cheswick Surveys) which is based in the outskirts of Nairobi, Kenya. Koome is a live wire with the use of geodetic GPS and total stations, as well as all the geometric calculations and software involved. When his eye is not behind a telescope, he loves to write articles about surveying. One of these articles was republished in an Australian magazine.

List of Acronyms and Abbreviations

2D	Two dimensional
3D	Three dimensional
ACSM	American Congress on Surveying and Mapping
ALTA	American Land Title Association
AR	Augmented reality
ASCE	American Society of Civil Engineers
ASPRS	American Society for Photogrammetry and Remote Sensing
AUSPOS	A free online GPS data processing service provided by Geoscience Australia
BLM	Bureau of Land Management
BM	Benchmark
CAD	Computer aided design
CADD	Computer aided design and drafting
CALTRANS	California Department of Transportation
CEP	Circular error probability
CORS	Continuously operating reference station
CSRS	Canadian spatial reference system
DEM	Digital elevation model
DGPS	Differential global positioning system
DInSAR	Differential interferometric synthetic aperture radar
DOP	Dilution of precision
DRMS	Distance root mean square
DTM	Digital terrain model
ECEF	Earth centered earth fixed
EDM	Electronic distance measuring
EEO	Equal employment opportunity
FAA	Federal Aviation Administration
FGCC	Federal Geodetic Control Committee
FGCS	Federal Geodetic Control Subcommittee
FGDC	Federal Geographic Data Committee
FIG	International Federation of Surveyors
GCP	Ground control point
GDOP	Geometric dilution of precision
GIS	Geographic information system
GPS	Global positioning system
GNSS	Global navigation satellite system
HDOP	Horizontal dilution of precision
IAG	International Association of Geodesy

ICSM	Inter-governmental Committee on Surveying and Mapping (Australia)
IGS	International GNSS Service
ITRF	International Terrestrial Reference Frame
MSL	Mean sea level
NAD83	North American Datum 1983
NGS	National Geodetic Survey
NMAS	National Map Accuracy Standards
NOAA	National Oceanic and Atmospheric Administration
NSRS	National Spatial Reference System
NSSDA	National Standard for Spatial Data Accuracy
OPUS	Online Positioning User Service
PC	Point of curvature (beginning of curve)
PDA	Personal digital assistant
PDOP	Position dilution of precision
PI	Point of intersection
PLSS	Public Land Survey System
PPM	Parts per million
PPP	Precise point positioning
PT	Point of tangency (end of curve)
QBS	Qualifications-based selection
RF	Representative fraction
RFP	Request for proposal
RINEX	Receiver Independent Exchange format
RMS	Root mean square
RMSE	Root mean square error
RTK	Real time kinematic
SBAS	Satellite based augmentation system
SPC	State plane coordinate
TDOP	Time dilution of precision
UAS	Unmanned aerial system
UAV	Unmanned aerial vehicle
USACE	US Army Corps of Engineers
USGS	US Geological Survey
UTM	Universal transverse mercator
VCV	Variance-covariance
VDOP	Vertical dilution of precision

Part I

Overview

1 Project Design Process

1.1 UNDERSTANDING PROJECT REQUIREMENTS

1.1.1 SCOPE, TIME, AND BUDGET

Project design, whether by *workflow* or *schematic*, requires an understanding of the project requirements. Almost everything significant that we do in life can be considered a project (e.g., buying a house, training to run a marathon, completing a degree at university, and job search), so we are all familiar with projects in general terms. *A project* is defined as a *temporary, unique endeavor undertaken to create a unique product, service, or result*. It is an exception, that is, it is not routine. Within a project, tasks are related to one another and to the end result of the project. A project has specific *goals* and *deadlines* and operates under the constraints or restrictions of *scope*, *time*, and *budget*.

The following is an example of a project:

> A company, AA Consultants, has been contracted by the City of Palm Desert to provide an up-to-date record of survey of the Highway 111 Corridor within the City of Palm Desert to be used as a baseline for future jurisdiction surveys. The project is to be completed by July 15, 2010 and has a contract value of $76,000.

Why is this considered to be a project? It is temporary, unique, and not a routine activity. It is based on a specific contract and has a defined end. It has a specific goal and the activities to be defined to reach that goal will be interrelated to produce the end result—the record of survey. The constraints of scope ("produce an up-to-date record of survey"), time ("to be completed by July 15, 2010"), and budget ("a contract value of $76,000") are defined.

These three constraints represent the essential elements of any project:

1. *Scope (what)* is the work of the project, leading to the product (result, outcome, service, deliverable, and performance).
2. *Time (when)* defines the schedule of the project, with start and end dates for the project as a whole, and the tasks and milestones.
3. *Cost (how much)* is defined by the resources used in the project (people, systems, equipment, data, and facilities).

Each of the three constraints is directly related to the other two (Figure 1.1). When they are carefully planned and managed accordingly, the project is considered "in balance," also known as "balancing the triad."

DOI: 10.1201/9781003297147-2   3

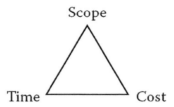

FIGURE 1.1   The triad of project constraints.

1.1.2 DESIGN FRAMEWORK

Project design is the basis upon which an approach to solving a problem is developed, together with the time and cost estimates (Figure 1.2). There can be a workflow design or a schematic design, or even a mix of both for the same project. A *workflow design* outlines the logical sequence or procedures that must be followed to accomplish the required goals. A *schematic design* includes the physical plan such as of a survey network. This book does not emphasize either of the two approaches because every project is different. In Part III of the book, we will discuss some important considerations of planning and design, which incorporate both the workflow processes and schema.

Before a project design can occur, some entity or client determines that a project is necessary and sends out a request to potential consultants. Interested consultants prepare proposals and submit them to the client. The client selects a winning proposal, and the work begins. At the heart of every proposal are the project design and a corresponding time and cost budget.

In the above example, a project design by the AA Consultants would have occurred somewhere between invitation to submit a proposal (i.e., prior to the award of the contract) and project initiation. This would entail, for instance, putting together a project plan by the AA Consultants, in line with the scope, schedule, and budget of the project work required to deliver desired results. On that basis, the AA Consultants submits a proposal to the *client*, who then evaluates the proposal and awards or denies the contract.

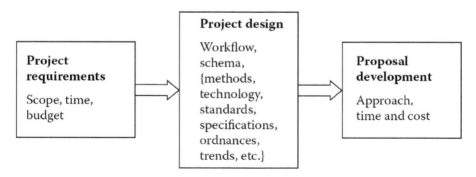

FIGURE 1.2   Project design framework.

Project design requires gathering, synthesizing, and analyzing information with enough objectivity and detail to support the project decision that makes optimum use of resources to achieve desired results.

1.2 INFORMATION GATHERING

A typical geomatics project involves gathering field information and measurements (field data is collected by ground surveys, aerial surveys, or by a combination of these two methods). The project information is used in locating, designing, and constructing civil infrastructure; establishing control for land boundary records and geographic information systems; mapping for engineering and land development; establishing baseline data for disaster monitoring; site feasibility studies; and so forth. Information gathering generally consists of (1) an examination of existing information about the project and (2) the physical gathering of ground information. Both information-gathering actions are of equal importance, and careful attention to detail during this process can often result in substantial savings in time and effort.

1.2.1 EXISTING SOURCES

Before any type of survey project occurs, perform a search for existing information. For the most part, the information described next can be obtained from government agencies. However, do not limit the search to these agencies. Much valuable information may be available from private consulting firms that have worked on similar projects. Sources of information that are helpful during the course of a survey may include survey control data, construction plans, existing photography, existing maps, plans and legal property descriptions, local landowners, and agency contacts.

1.2.1.1 Survey Control Data

Horizontal and vertical control are crucial to performing an accurate and correct survey. Wherever practically possible, base the survey on horizontal coordinates and vertical elevations from established control points (e.g., in the United States, use first-order or second-order National Geographic Survey [NGS] control points). The horizontal and vertical control information can be obtained by contacting agencies, such as NGS, directly or by doing research using a variety of Internet resources, such as Google.

Here are examples of the control information to gather:

Horizontal control: Monument name, location (state, country, etc.), year the monument was established, coordinates (geodetic or plane), coordinate system (e.g., SPC, UTM), geodetic datum (e.g., NAD83, ITRF), order of accuracy, station recovery/condition notes, azimuths and distances to neighboring monuments, and any other pertinent information.

Vertical control: Monument name, location (state, country, etc.), elevation (in feet or meters), order of accuracy, date established and by whom, station recovery/condition notes, and any other pertinent information.

1.2.1.2   Existing Photography

The use of photography as a source of preliminary project information is somewhat limited. General project layouts can usually be obtained from readily available maps rather than from photographs. On the other hand, existing aerial photographs for a current project can often be used by the photogrammetric engineer. If the control points that are referenced in the photography can be reestablished by a ground survey, the photographs may be usable. The existence of old aerial photographs often indicates the presence of aerial maps. In construction projects such as highways, final construction reports may also be a source of helpful photographs. Aerial photographs are usually available from agencies or private consulting firms. However, there is also vast amount of information on the Web that could lead to various other sources of photographic information, including commercial. So, you could easily browse your way into existing digitized photographs or even LIDAR data for height information just by a stroke of a few keywords.

1.2.1.3   Existing Maps

If the project is in the United States, you generally have 7½ or 15-minute quadrangle (topographic) maps available covering the desired project limits. These maps are available from both the U.S. Geological Survey (USGS) offices or its Web site, and from many private vendors for a minimal fee. They provide a wide variety of control and terrain information.

For most types of geomatics projects, there exist a variety of available maps. By using these maps, much of the field gathering of information can be reduced.

Generally, whether the project is located in the United States or another country, you can do an Internet search using the keywords from information provided under each agency listed next to locate the availability of existing maps for the project area. A list of agencies that provide maps containing survey information within the United States includes:

USGS: The USGS provides access to quads, topographic, and index maps; benchmark locations, level data, and table of elevations; stream flow data and water resources; geologic maps; horizontal control data; monument locations; seismological studies; and aeronautical and magnetic charts.

NGS: The NGS provides access to topographic maps; coastline charts; benchmark locations, level data and table of elevations; horizontal control data; state plane and UTM coordinates; and monument locations.

Bureau of Land Management: The Bureau of Land Management provides access to township plots, showing land divisions; and state maps, showing public lands and reservations.

Department of the Army: The Department of the Army provides access to topographic maps and charts, aeronautical charts, and hydraulic and flood control information.

Department of Transportation: The Department of Transportation provides access to easements, right-of-way maps, and permits for bridges in navigable rivers.

Department of Agriculture: The Department of Agriculture provides access to soil charts, maps, and maps; and forest resource maps including topography, culture, and vegetation classification.

Postal Service: The U.S. Postal Service provides access to delivery maps by counties (showing rural roads, streams, etc.).

Local governments (state, county, city): The local governments provide access to street and zoning maps; drainage and utility maps; horizontal and vertical control data.

1.2.1.4 Plans and Legal Property Descriptions

As-constructed plans can be an excellent source of preliminary information, especially if dealing with a civil infrastructure project such as roads, highways, or bridges. Depending on the composition of the construction plans, a surveyor may obtain the position and condition of existing control points, right-of-way monuments, benchmarks, and construction monuments. If it is desirable to use the existing centerline stationing and location, the centerline control points (such as PC and PT) can be obtained from the as-constructed plans. The horizontal alignment information is also often obtained from these plans. Other information available from as-constructed plans includes the types and location of drainage systems, structures and special features, property descriptions, and boundary lines.

Legal property descriptions, survey records, and reports provide information concerning the identity and location of property corners. Ties to property corners, which are commonly carried out during cadastral surveys, are also useful for route design and right-of-way projects.

1.2.1.5 Agency Contacts and Interviews

Before any surveying activity begins on a project, contact the local representatives of any concerned agency (stakeholder). The agency contact may be able to provide additional information about the availability of existing survey data and the type of ground survey that may be appropriate. The second purpose of contact is to inform the agency that a survey is about to be performed, and for this, briefly describing the intended surveying activities is necessary.

The contact also provides a means to interview the stakeholders on any special requirements or restrictions, such as limitations on cutting vegetation, noise requirements, property access permissions, environmental restrictions, recreational uses, scenic routes, and so forth.

Affected property owners should also be contacted. A letter to the property owner asking permission to enter property for survey work is recommended. Retain any signed documents for the project records. Where contact cannot be made or permission granted, try other ways rather than trespassing.

1.2.2 GROUND INFORMATION

The type of survey that can be used to gather ground information for a project area can be divided into two categories: reconnaissance and preliminary field surveys, and the actual field surveys.

1.2.2.1 Reconnaissance

A reconnaissance survey is the examination of a large area to determine the overall feasibility of the fieldwork portion of a project. The following are some of the many goals of such a preliminary survey:

- To assess the accessibility of the project area
- To assess the existing project controls and their conditions
- To assess the feasible or alternative project points
- To assess the feasible or alternative project routes
- To assess the feasible or alternative field methods or techniques
- To assess environmental conditions such as existence of a wetland
- To assess the intervisibility of desired project points
- To assess the sky visibility at desired project points.

Aerial photographs, maps, and images acquired from a preliminary research of the project area are often useful. In rare cases, it may be necessary to carry out preliminary field surveys to gather planning data for the main survey.

The evaluation of feasible alternatives (i.e., a comparison of the project design, e.g., in terms of the project point locations, data collection methodologies, and alternatives) in sufficient detail is necessary to decide the most feasible cost-effective solution.

1.2.2.2 Surveys

The types of information gathered during field surveys can be divided into three categories: planimetric, topographic, and cadastral.

1. *Planimetric:* Planimetric data consists of natural and political boundaries, natural vegetation, and cultural items such as signposts, trees, and buildings. Using *ground surveying* techniques, these items are located relative to survey control monuments. Specific items are surveyed with side shot measurements taken from these control points. Only the horizontal positioning (coordinates) for each point is required to plot the item on a planimetric map. However, when using total station surveying equipment, it is recommended that the elevation of each point be obtained. This additional data aids the plotting of contour intervals during the topographic mapping process.

2. *Topographic:* Topographic information gathering begins where planimetric information leaves off and consists of obtaining horizontal coordinates and vertical elevations of ground points. The intent of topographic data gathering is to obtain enough ground points to accurately describe the general relief of a specific area.

   There are three methods of mapping a given area with topographic shots. The first is to use *alignments and cross sections.* An alignment is usually a straight line connecting ground control points. For such a line, you can establish points at given intervals, say 20 m, with the spacing of these points generally based on the type of land features and relief along the route or as otherwise guided by the project requirements. Cross sections are taken perpendicular to the alignment at these regular intervals, and all the points in the project essentially form a grid of coordinates that can be used to construct a contour map.

The second method is the use of *radial surveying*. The instrument is set up on a point with known elevation, and coordinates and readings are taken in a radial pattern around the instrument. Major break(line)s in the terrain (such as edges of shoulders, catch points, and drainages) are usually strung together in a series of sequential shots. These data points are called breaklines (or discontinuities) and are treated differently from other random shots. A general description of the terrain can then be obtained, using a digital terrain model (DTM) to build an accurate contour map.

The third method is the use of *aerial photography and photogrammetric techniques* such as LIDAR to plot topographic data.

3. *Cadastral:* A cadastral survey is used to locate property boundaries and monuments and determine the respective coordinates. This information may be obtained disregarding elevation. Because property and right-of-way documents are often based on the actual location of cadastral monuments, the points can be verified by running traverses through them or by using the mean of two independent side shots.

1.3 DESIGN APPROACHES

1.3.1 Workflow Design

Careful planning at the beginning of a project will help you avoid hours of unnecessary work and redundant tasks. The following basic steps can be followed to carry out a geomatics project:

1. *Project goals:* What is the purpose of the project? What is the spatial extent (and ground resolution) of the study? What type of data do you need to achieve your goals? What are the sources of these data, and what are the appropriate types of data to answer the project questions?

2. *Methodology:* Constructing a logical flow sequence that details the project steps will make the success of the project more likely. What types of procedures and analyses will you perform? A project plan should include: (1) an outline of procedures required for data collection or gathering, (2) a logical sequence of procedures to be performed, and (3) a list of all the information and data required for each step.

3. *Data and resources:* Before you embark on the project, you should do an inventory of the data requirements and sources of information. Even with the widespread availability of digital data on the Internet, many projects still require data collection, input, and integration. For instance, in a Geographic Information System (GIS) project, check if the data are already in digital format or whether you have to scan paper maps or input data from other sources. What software systems and equipment are required, and are they available?

4. *Analysis:* What is the measure of confidence in the project? Often you will find that once a project is started, there is a need to revise the procedures originally intended. A preanalysis may be necessary to control the project in terms of time, cost, and accuracy constraints. In addition, once the data collection and analysis are complete, you should evaluate the accuracy and validity of the results. If applicable, a repeat of fieldwork may be required.

5. *Presentation:* Results should be presented in a format suitable for the client, organization, or the audience, such as PowerPoint presentation, journal paper, written reports, field notes, maps, GIS system, CADD files and drawings, and other digital media.

1.3.2 SCHEMATIC DESIGN

A *schematic design* and *preanalysis* will allow experimentation with different variables so as to meet or exceed the project (accuracy) requirements.

Case in point: What are the benefits of network design for a GNSS survey, given that the accuracies of individual GNSS baselines are a function of satellite geometry and not survey network geometry?

The short answer is that network design helps to provide a measure of confidence in the planned survey before you enter the field. That measure of confidence is a function of the network design. The design variables with which you can experiment a GNSS survey network design include: (1) the number and physical location of survey points, (2) the number and types of observations to be measured, and (3) the observation standard deviations (standard errors) you expect to achieve in the field. Altering any one of these variables will change the estimated confidence of the survey project. Network design allows you to perform what-if analysis on these variables so that you can estimate how you will do in the field.

A preanalysis of GNSS survey network design will help you achieve the following project design goals:

1. *Performing the project in a cost-effective way:* Can the survey be performed with fewer points on the ground, while still meeting accuracy requirements? Further, if you could select locations on the ground that was easy to gain access to and make observations from, and still be able to meet accuracy requirements, wouldn't that be beneficial?

2. *Determination of the field procedures and equipment needed to achieve accuracy requirements:* This could be something as simple as using a more accurate total station or perhaps changing your field procedures a bit to achieve better accuracy (e.g., making terrestrial measurements during the cooler times of day, better instrument/target setups, making additional measurements, etc.).

3. *Determination of whether you should take on the project:* Based on the accuracy requirements, you may decide that given the nature of your equipment and/or crew, you may not be able to meet the requirements and therefore should pass on the project.

Network design allows you to achieve these goals by providing you with estimates of the accuracy that will be achieved given the input observation types, their standard deviations, and station locations in the survey. After an initial design, you may discover that the accuracy estimated will not meet the survey requirements. Using an iterative process of changing out the variables, you may find a way to satisfy the accuracy requirements.

Before bidding on a new project, you might initially set up an elaborate design with many different observation types built in. After running the design and satisfying the confidence requirements, you might then scale back the network with fewer stations and observations. After running the design again, you may happily discover that you are still within the accuracy requirements of the project, but now the project will cost less to perform.

Next, you might consider using only GNSS for the project. However, after running your proposed network through the design process, you might discover that a problem has emerged that cannot be fixed through GNSS alone. In fact, you may need to add terrestrial observations for some portion of the project in order to stay within accuracy requirements. This might occur in an area in which you have poor satellite visibility or in an area in which the points you need to establish are only a few hundred meters apart. Perhaps only the terrestrial equipment can give you the accuracy you need in these areas.

After the design is completed, you will have created a blueprint for the field crew. That blueprint will tell them roughly where to locate the stations, the types of observations to measure at each station, and the level of accuracy needed for those observations. You could conceivably use GNSS in one section of the project, a 10-second total station in another section, and a 1- to 2-second total station in yet another section of the project. Through the use of network design, you can determine how the survey should proceed.

The most important element is achieving "in the field" what you designed in the office. If you are unable to measure angles to ± 5 seconds or measure distances to ± 0.004 m (as specified in the design), then your project will probably not meet the expectations derived from design. Bottom line: Don't be overly optimistic about what you can achieve in the field.

1.4   SCHEDULING AND COST ESTIMATING

1.4.1   GENERAL STEPS IN GEOMATICS PROJECTS

In scheduling of geomatics projects, it is helpful to first understand the overall process in completing a specific project. Here we will look at five case examples. But first, let us summarize the common types of geomatics projects.

> *GNSS survey:* GNSS surveys use portable receiving antennas to gather data transmitted from satellites, which are used to calculate the position of the object being located on the surface of the earth. The receiving antennas can be miles apart and still obtain very accurate data. GNSS surveys are used to establish coordinate control points for projects such as for State Plane

Coordinate Systems, large boundary surveys, and subdivision surveys. They can also be used to collect data for GIS/Land Information Systems, such as the location of streets, homes, businesses, electric, phone and gas utilities, water and sewer systems, property lines, soil and vegetation types, water, and courses. This data can be used in future planning, preservation, and development.

Topographic survey: A survey locating improvements and topographic features such as elevations of the land, embankments, contours, water courses, roads, ditches, and utilities. This survey can be used in conjunction with a location survey in order to prepare a site design map, a subdivision map, or an erosion control plan.

Boundary survey: A survey of the boundary of property according to the description in the recorded deed. Interior improvements, such as buildings and drives, are not located. Any improvements along the boundary affecting the use of or title to the property are located, such as fences, drives, utilities, buildings, sheds, and streets. Missing corner markers are replaced. A map showing the boundaries and improvements along the boundaries is prepared.

Location survey: A boundary survey with the additional location of all the interior improvements. Missing corner markers are replaced. A map showing the boundaries and improvements is prepared. This type of survey may be required for the acquisition of a loan.

Site planning survey: This survey uses a boundary and topographic survey as a base to design future improvements. It can be a design for a house, a residential subdivision, a store, a shopping center, a new street or highway, a playground, or anything else.

Subdivision survey: This often includes a topographic survey of a parcel of land, which will be divided into two or more smaller tracts, lots, or estate division. This can also be used for site design of lots, streets, and drainage. It is for construction and recording.

Construction survey: Using surveying techniques to stake out buildings, roads, walls, utilities, and so forth. This includes horizontal and vertical grading, slope staking, and final as-built surveys.

ALTA/ACSM survey: This is a very detailed survey (mainly in the United States) often required by lending institutions. The request for this survey must be in writing and be included with all of the deeds and easements affecting the property, along with the deeds to adjoining properties. A list of items to be located as noted in the ALTA/ACSM publication can be included.

The typical steps to be taken during a particular project can be defined on the basis of the type of project, a subdiscipline or technology focus. This will be illustrated by the following five case examples.

1.4.1.1   Steps in a GNSS Control Survey Project

1. Determine the scope of the project.
2. Determine project requirements (accuracy, number of stations, spacing, etc.) from both the scope and relevant standards.

3. Research station information for existing horizontal and vertical control stations.
4. Determine suitability of existing control for GNSS observations and select those stations that are required to meet standards for the project.
5. Select sites for new project stations ensuring clear access to satellite signals and no multipath problems, set new survey monuments, and prepare station descriptions.
6. Design the project layout (network).
7. Determine number and type of receivers (single or dual frequency; P-code, C/A-code, etc.) required to meet project specifications.
8. Plan observation schedules (station observation time accounting for satellite availability, ensure redundancy, etc.).
9. Conduct GNSS observations and complete observation log.
10. Download data from GNSS receiver(s) and make backup copy.
11. Receive and process data at central processing location.
12. Review all data for completeness.
13. Perform minimal constrained adjustment.
14. Review results for problem vectors or outliers.
15. Reobserve problem vector lines, if necessary.
16. Perform constrained adjustment and review results.
17. Incorporate precise ephemeris data if appropriate.
18. Prepare final report with all sketches, maps, schedules, stations held fixed (including coordinates and elevations used), software packages used, station description, final adjustment report, and list of coordinates.

1.4.1.2   Steps in a Topographic Mapping Project

1. Determine the scope of the project.
2. Determine scale and accuracy requirements (map scale, contour interval, product resolution/accuracy, etc.).
3. Research project information (existing control, satellite imagery, aerial photos, LIDAR data) and available resources (equipment, personnel).
4. Design the project layout and decide the data collection methods that can meet project requirements.
5. Determine number and observation types required to meet project specifications, if applicable.
6. Plan observation schedules, if applicable.
7. Conduct observations (topo/aerial survey(s)) and/or gather project data (satellite imagery, digital orthophotos, LIDAR data, DTM/DEM, etc.).
8. Receive and process data at central processing location.
9. Perform aerial photo triangulation, if applicable.
10. Perform digital photo rectification, if applicable.
11. Process and review all data for completeness.
12. Generate map and contours and check map for correctness and completeness.
13. Correct map and/or reobserve problem areas, if necessary.
14. Prepare final map and a final report with all pertinent sketches, schedules, stations used, software packages used, and list of coordinates.

1.4.1.3   Steps in a Boundary Survey Project

1. Determine the scope of the project.
2. Research project information (i.e., search existing survey records, titles, notes, descriptions, maps, photos, and any other pertinent data).
3. Design the project layout and decide the data collection methods that can meet the boundary control requirements.
4. Plan observation schedules, if applicable.
5. Conduct field observations.
6. Receive and process data at central processing location.
7. Compute locations of missing boundary monuments.
8. Review all data for completeness and accuracy.
9. Reobserve problem areas, if necessary.
10. Prepare map with notes and descriptions (including coordinates, distances, bearings) and a final report with all pertinent notes, sketches, schedules, stations used, and software packages used.

1.4.1.4   Steps in a GIS Project

1. Determine the scope of the project.
2. Identify project goals: What is the purpose of the project? What is the research question? What is the spatial extent and ground resolution of the project? For example, if a soils map is the answer to the question, what is the spatial distribution of soils in the locality?
3. Identify the data and information needs. What type of spatial data do you need to achieve the project goals? What are the sources of these data, and what are the appropriate types of data to answer these questions?
4. Determine data accuracy requirements.
5. Identify sources of existing data and information.
6. Perform an inventory of data and sources of information. There is widespread availability of digital data; however, GIS projects are still essentially about data collection, input, and integration. Check if the available data are already in digital format. Will you have to scan paper maps or input data from other sources? Will fieldwork be necessary?
7. If applicable, identify target features that will be located and how they will be located. The project goals and objectives should guide the identification of target features and information to be gathered about the target features, as well as ideal ways to represent or symbolize them (e.g., point, line, and polygon) for GIS analysis.
8. Select or design the methodology for data collection, and data analysis and integration. What types of data analyses will you perform? Overlays? Statistical regressions? Spatial interpolations?
9. Collect data or carry out field observations as applicable.
10. Process and/or compile input data for the GIS system.
11. Analyze GIS input data.
12. Evaluate accuracy and validity of results.
13. Revise procedures, if practical and necessary.

14. Prepare results in a suitable format (e.g., digital, paper, or other media) and a final report on procedures and software packages used.

1.4.1.5 Steps in Aerial Survey/Mapping Project

1. Determine the scope (scale and location) of the project.
2. Carry out mission planning, such as to determine aircraft type (e.g., the most suitable unmanned aerial system [UAS]), define flight parameters, and check the local weather forecasts (predictions) for the project's location.
3. If using a UAS/drone, research information on pertinent local laws and regulations.
4. Define ground control points for georeferencing the aerial survey/mapping.
5. Determine processing software application and workflow.
6. Carry out the flight mission.
7. Download and process data.
8. Prepare results in a suitable format for deliverables such as dense point clouds, digital elevation models, orthomosaic, and contours.

1.4.2 PROJECT SCHEDULING

A project schedule is necessary so that a provisional budget can be developed. It is a plan of activities (milestones) and their timeframe. In other words, it is concerned with *when* things occur over the course of a project. It includes the processes required to ensure timely completion of the project. Figure 1.3 shows an illustration of a project schedule using a Gantt chart.

Every project must have at least one deadline, and usually there are more deadlines imposed during the course of the project based on specific tasks that are required to complete the project. The tasks are placed in the order in which they will be carried out, with interdependent tasks properly planned out. It is also important that the project objectives have been accounted for (e.g., the client's completion date) and suitable resources (people and equipment) are available for the planned tasks.

Project scheduling also serves as a check on the project's viability. If it cannot be completed successfully to meet the client's deadline(s), for example, a renegotiation of scope or schedule or budget may be required. In addition, missing tasks that were originally overlooked may be identified during the project scheduling stage.

Although a project schedule can be prepared by one person, it is more effective if it is developed with a project team when possible. The person who understands each task or a set of tasks best will have the best understanding of the sequence in which the tasks should be performed and the best estimate of the duration of each task. These estimates can then be calibrated by the project leader or manager.

Sometimes it is necessary for schedule planning to enlist the help of people other than the project team. Subcontractors can provide their own estimates, but it may be necessary to negotiate with them in order to meet the required schedule. Other managers within the organization or external experts can provide input, particularly if they have worked on similar projects.

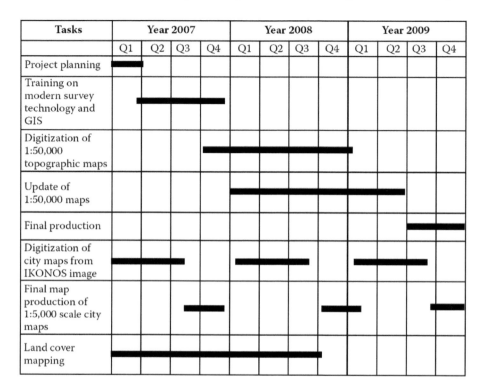

Tasks	Year 2007				Year 2008				Year 2009			
	Q1	Q2	Q3	Q4	Q1	Q2	Q3	Q4	Q1	Q2	Q3	Q4
Project planning												
Training on modern survey technology and GIS												
Digitization of 1:50,000 topographic maps												
Update of 1:50,000 maps												
Final production												
Digitization of city maps from IKONOS image												
Final map production of 1:5,000 scale city maps												
Land cover mapping												

FIGURE 1.3   A project schedule showing timeframe of interrelated activities.

The first step in developing a project schedule is to list all the specific tasks to be performed in the project to produce the required deliverables. The next logical steps include: (1) identify predecessors for all the tasks, (2) estimate durations (work periods to complete) for all tasks, (3) identify any intermediate and final dates to be met (constraints), (4) identify all activities outside the project that will affect the performance, and (5) put all the tasks on a time scale (Gantt chart).

1.4.3   COST ESTIMATING PRINCIPLES

Cost estimating is concerned with *how much* the project will cost to complete. The prerequisites are that project resources are planned in a timeframe as explained in the previous section. Having developed a project schedule, resource costs are calculated to produce a cost estimate for the project. The cost estimate can subsequently be used to develop a cost budget.

The three general stages of cost estimating that a project manager needs to end up with a baseline cost budget are as follows:

1. *Plan project resources:* Determine what resources (people, equipment, materials) are needed in what quantities to execute the project tasks.
2. *Estimate resource costs:* Develop an approximation of the cost of the resources needed to complete the project tasks to produce a cost estimate.

3. *Budget costs:* Allocate the resource costs to the project tasks over the length of the project to produce a cost budget.

Information on project resources can be obtained from the scope statement, the list and description of tasks and their estimated durations, and ultimately the project schedule. Another source of information could be the organization's archive of past projects; files on similar past projects should have good data on who and what were used to accomplish similar tasks and what they cost. In some organizations, there could also be a "resource pool" from which relevant skills for the project could be identified. Sometimes experts could come from other organizations. At this stage, it is also important to ensure that policies and procedures of the project host organization are taken into account in resource planning. Such policies include, for example, policies related to length and type of work week and work day, holidays and vacations, and hiring of consultants or contractors.

Once the resource items are identified, their costs can be estimated. Typically, the largest cost item is labor (i.e., the people who will be doing the project work). The amount of time each member will be spending on each task must be determined, the unit cost figured out, and the cost of labor totaled. Other resources whose expenses are applied to the project cost include equipment, materials, travel and living, subcontracts, training, and so forth. When doing the project costing it is also important to distinguish between the *direct costs* and the *indirect costs*.

Direct (or variable) costs can include

1. *Labor:* The cost of the time of the people who will work on the project
2. *Specialized systems and software:* The cost of systems or software purchased for the project or time-based charge for their usage
3. *Equipment:* The cost of tools or equipment purchased for the project or a time-based charge for equipment use
4. *Materials and supplies:* The cost of materials used on the project (e.g., monumentation, plotter, and paper)
5. *Travel and living:* The cost of travel carried specifically for the project (e.g., travel to and from a field site in terms of vehicle mileage, airfares, and cost of accommodation and meals while in the field)
6. *Subcontracts:* The cost of subcontracts for completing project work
7. *Fees:* Fees charged specifically for work on the project (legal fees, financial fees, agent fees for international work, title search fees, etc.)
8. Courier, postage, and freight costs for the project.

Indirect costs are the costs that do not specifically relate to a particular project. Sometimes referred to as *fixed costs* or *overhead costs*, they represent the costs of operating a business that provides the services for the project. These costs are shared among all projects that are carried out in an organization. They may include

1. *Facilities:* The cost of providing the physical location for carrying out project work, and cost of shared resources used for such operations. Examples include office space, telephone, computer systems, equipment repair and

maintenance, Internet access, journal subscriptions, and professional training.

2. *Overhead labor:* Administration costs, human resources, marketing/sales, and other staff who support the project, but are not directly charged to the project.

3. Other requirements specific to the project location such as taxes.

Indirect costs are allocated to projects in many different ways, often on a percentage basis depending on the size of the project or using some other criteria. Most organizations would have a standard policy on how this is done.

Finally, the budget cost estimate will incorporate a risk assessment for the project. Based on the risk assessment, budget *contingencies* can be applied to allow for some flexibility in budget management when and if problems occur during project execution. A common practice is that some managers will include a contingency of, say, 10%, on every project. This practice has its drawbacks, for instance, a tendency to manage to the limit of the total budget, rather than to the budget as planned without contingency percentage added. Another drawback is that the business might gain a reputation for always overestimating the budget. For proper costing, a list of potential problems and their impact on the project can be outlined to justify budget contingencies.

A cost budget is prepared based on the cost estimate of all resources required to complete the project. A *cost budget* is a detailed, time-phased estimate of the costs of all the resources required to perform the project work over the entire duration. In other words, it takes the cost estimate and spreads it over the budget schedule, based on the timing of the project tasks.

1.5   WRITING PROPOSALS

Each project should be preceded by a detailed description of what is to be accomplished, together with a proposal or estimate of the time and cost required.

(Morse and Babcock, 2007, 325)

Businesses, small or large, customarily respond to RFP (request for proposal) to win projects for their survival. In that process, it is important to understand the customer's problem and the elements of strategy that make a winning proposal. The steps to developing a proposal include (1) understanding the project requirements; (2) planning, intelligence gathering, and design; (3) scheduling and cost estimating; and (4) writing the proposal.

A well-written proposal should have the following attributes: (1) evidence of a clear understanding of the project (client's problem); (2) an approach, program plan, or design that appears to the client well suited to solving the problem and likely to produce desired results; (3) convincing evidence of qualifications and capability to carry out the project; (4) convincing evidence of dependability as a consultant or contractor; and (5) a compelling reason to be selected (i.e., a winning strategy).

Understanding the project (client's problem) is the key to writing a successful proposal. Some firms strategically identify new opportunities long before an RFP is issued. They prepare for new opportunities and estimate the resources and capabilities that will be required to meet expected future needs of potential clients. Such preparation may include, for example, developing the necessary technical skills and acquiring other needed resources in advance.

Having identified or received (and reviewed) an RFP, a bid or no-bid decision is made based on the understanding of the requirements of the project, and the capabilities of the firm and those of the competitors. A preanalysis of design (incorporating the budgetary constraints) can be applied to decide whether to bid or not to bid.

If a decision is made to bid, then the proposal preparation must pay very close attention to the language of the RFP. An RFP typically includes a cover letter, a statement of *scope of work* (which specifies work to be performed), the required schedule, specification of the length and content desired in the proposal, any pertinent standards and specifications, and the required deliverables.

In response to an RFP, a written proposal should include an executive summary, the capabilities (or strengths) of the firm, work schedule and a cost budget, and project deliverables. The proposal should project a professional image as much as possible—exercise proper writing skills, include graphics where necessary, and consider the legal aspects of the project. However, *do not do the whole project during the proposal!*

In general, the expected contents of a proposal will include

1. A management proposal discussing the company, its organization, its relevant experience, and the people proposed to lead the project
2. The technical proposal outlining the design concept proposed to meet the client's needs
3. The cost proposal including a detailed cost breakdown, but often also discussing inflation, contingencies, and contract change procedures.

In Chapter 12, we will look at further details on writing geomatics proposals.

BIBLIOGRAPHY

Coleman, D. 2007. *Manage Your GeoProject Effectively: A Step-by-Step Guide to Geomatics Project Management*. Calgary, Alberta, CA: EO Services.

Crawford, W. G. 2002. *Construction Surveying and Layout: A Step-By-Step Field Engineering Methods Manual*. 3rd ed. Canton, MI: Creative Construction Publishing.

Ghilani, C. D. and P. R. Wolf. 2008. *Elementary Surveying: An Introduction to Geomatics*. 12th ed. Upper Saddle River, NJ: Prentice Hall.

Morse, L. C. and D. L. Babcock. 2007. *Managing Engineering and Technology*. 4th ed. Upper Saddle River, NJ: Prentice Hall.

North Carolina Society of Surveyors. 2009. *Facts about having your Land Surveyed*. http://www.ncsurveyors.com (accessed July 17, 2009).

Robillard, W. G., D. A. Wilson and C. M. Brown. 2009. *Brown's Boundary Control and Legal Principles*. 6th ed. New York: Wiley.

Van Sickle, J. 2007. *GPS for Land Surveyors*. 3rd ed. Boca Raton, FL: CRC Press.

EXERCISES

1. Explain the following terms and phrases:
    i. Project scope
    ii. Project constraints
    iii. "Balancing the triad"
2. Which of the following is not a survey network design goal?
    a. Performing survey project in a cost-effective way
    b. Determination of the field procedures and equipment needed to achieve accuracy requirements
    c. Determination of whether you should take on the project
    d. To gain experience for future tasks
3. Network design (e.g., for a GNSS survey) allows for experimentation with different variables (such as point locations, observation types, and expected accuracies) to estimate the "confidence" of a survey project. It would be most appropriate to carry out this important task:
    a. Before bidding on a project
    b. After completion of the fieldwork, but prior to office computations
    c. After learning that the project is going to cost more than you intended
    d. Before submitting the final reports of the project
4. Designing a pipeline with minimum cost. This question involves using geometry and differential calculus. It requires that you determine the most cost-effective pipeline route in connecting various wells in an oil fertile area. Figure 1.4 shows a schematic view of the wetland and the corresponding simplified rectangular model. In this problem, we are connecting a pipeline from a well at point A to another well at point B. Costs are associated with material (cost of pipe of $1.50/foot) and terrain type (normal terrain

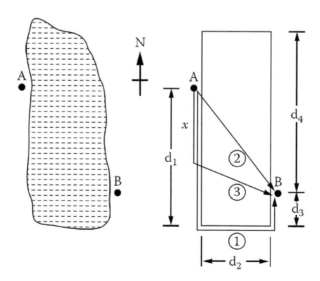

FIGURE 1.4   Plan view of the wetland pipeline project.

installation cost of $1.20 per foot). Installation in the wetland requires an additional track hoe at a cost of $60/hour. In a 10-hour day, the track hoe can dig approximately 300 feet of trench, and thus there is an additional cost of:

$$\frac{\$60/\text{hr}}{30\text{ft/hr}} = \$2/\text{ft}. \tag{1.1}$$

This gives a wetland installation cost of $4.70 per foot. Three routes are considered (labeled 1, 2, and 3 in Figure 1.4):

Route 1: $\text{Cost} = 2.7(d_1 + d_2 + d_3)$

Route 2: $\text{Cost} = 4.7\sqrt{(d_1 - d_3)^2 + d_2^2}$

Route 3: $\text{Cost} = 2.7x + 4.7\sqrt{(d_1 - d_3 - x)^2 + d_2^2}, \qquad 0 \le x \le d_1 - d_3$

A possible solution for the third route is to *optimize the cost function on a closed interval*, that is, find the derivative of the cost function and subsequently find the global minimum of the derivative. This can be done by setting the derivative equal to zero and solving for x.

Assume the following values for the coordinates of wells at points A and B (Table 1.1) and the rectangular model representing the wetland (Table 1.2):

Suggested solution procedure:

i. Compute the dimensions of the rectangular model in feet
ii. Compute numerical values for d_1, d_2, and d_3 of the pipeline route
iii. Calculate the optimum value for x
iv. Calculate the global minimum cost for Route 3
v. Find the most cost-effective route. Is it any of the following?
    a. Route 1 costing approximately $89,100.00
    b. Route 2 costing approximately $78,223.59
    c. Route 3 costing approximately $80,100.00

TABLE 1.1

Coordinates (in feet) of Wells at Points A and B

Point	N	E
A	62,000	45,000
B	48,000	54,000

TABLE 1.2

Rectangular Model (in feet)

N	E
70,000	45,000
43,000	45,000
70,000	54,000
43,000	54,000

5. If the coordinate information and the rectangular model for the wetland in Figure 1.4 were not available, and the client asked you to provide them using data from actual field surveying procedures, what specific questions would you ask to define the scope for such a task? Having defined the scope, what existing information sources would you consider? What surveying and mapping methods could you use to collect field information, if necessary? Give reasons for your answers.

Part II

Contemporary Issues

2 Standards and Specifications

2.1 DEFINITIONS

A standard attempts to define the quality of the work in a way that is ideally independent of the equipment or technology in use. A specification describes how to achieve a certain standard with a given set of tools, equipment, or technologies (Craig and Wahl, 2003, 93).

The following are typical definitions of *a standard*:

1. An exact value, or concept thereof, established by authority, custom, or common consent, to serve as a rule or basis of comparison in measuring quantity, content, extent, value, quality, and capacity.
2. The type, model, or example commonly or generally accepted or adhered to; criterion set for the establishment of a practice or procedure.
3. The minimum accuracies deemed necessary to meet specific objectives; a reasonably accepted error; a level of precision of closure; a numerical limit on the uncertainty of coordinates.

Specifications are the field operations or procedures required to meet a particular standard, the specified precision and allowable tolerances for data collection and/ or application, the limitations of the geometric form of acceptable network figures, monumentation, and description of points.

2.2 APPLICATION MODES OF A STANDARD

Craig and Wahl (2003, 94) identify three application modes of a standard:

> A standard can be applied as a design tool, a requirement, and an evaluation tool. Used as a *design tool*, a standard will enable us to assess what equipment and methods we need to use on a particular project in order to achieve the standard. This application is part of planning for new work. If viewed as a *requirement*, a standard is applied during the duration of the project to ensure that the work complies with stipulated quality requirements. And lastly a standard can be applied as an *evaluation tool* to work of any source and vintage in order to classify the work so that various users can make best and proper use of the data from that source for varying purposes.

DOI: 10.1201/9781003297147-4

2.3 UNITS OF MEASURE

The third definition of a standard (Section 2.1) is the most precise for geomatics applications. Standards are interpreted numerically in units of measure, and although there are many instances in which ratios and percentages are used to define standards, their interpretation is in units of measurements. Outlined in the following sections are the common metric scales (Table 2.1), selected conversion factors for both metric and imperial scales (Tables 2.2–2.4), and the common mapping scale terminologies. Metric units are nowadays widely used worldwide, but the imperial units (miles or yards) are also still common in some countries such as the United Kingdom and the United States. Metric units are widely used in science and industry.

TABLE 2.1
Metric Units

Millimeter (mm) = 0.001 m
Centimeter (cm) = 0.010 m
Decimeter (dm) = 0.100 m

Meter (m) = 1.000 m
Decameter (dam) = 10.000 m
Hectometer (hm) = 100.000 m
Kilometer (km) = 1000.000 m

TABLE 2.2
Selected Units of Linear and Square Measure

	Feet	
	The short form is ft or (')	
1 Imperial Foot	= 0.30479947 m	
1 International Foot	= 0.30480000 m	
1 U.S. Survey Foot	= 0.30480060 m	
1 U.S. Survey Foot	$= \dfrac{1200}{3937}$ m	
1 Indian Foot	= 0.30479841 m	
	Inch	
	The short form is in or (")	
1 Inch (")	= 1/12th (international) foot = 0.0254 m	
	Yard	
1 Yard	= 3 (international) feet = 0.9144 m	
	Mile	
1 Statute Mile	= 5,280 (international) feet	
1 International Nautical Mile	= 6,076.10 ft[a] = 1,852 m[a]	
1 Meter (m)	= 39.37 inches (in)	

(Continued)

TABLE 2.2 (*Continued*)
Selected Units of Linear and Square Measure

	Feet
1 Kilometer (km)	= 0.62137 miles
1 Mile	= 80 ch = 1,760 yards
1 mm²	= 0.00155 in²
1 m²	= 10.76 ft²
1 km²	= 247.1 acres
1 hectare (ha)	= 2.471 acres
1 acre	= 43,560 ft²
1 acre	= 10 ch², i.e., 10 (66 × 66 ft)
1 acre	= 4,046.9 m²
1 ft²	= 0.09290 m²
1 ft²	= 144 in²
1 in²	= 6.452 cm²
1 mile²	= 640 acres (normal section, U.S.)

[a] This distance is a function of the spheroid in use and will vary.

TABLE 2.3
Common Angular Units of Measure

1 revolution	= 360° = 2π radians
1° (degree)	= 60′ (minutes)
1′	= 60″ (seconds)
1°	= 0.017453292 radians
1 radian	= 57.29577951° = 57°17′44.806″
1 radian	= 206,264.8062″
1 revolution	= 400 grads (also called gons)
tan 1″	= sin 1″ = 0.000004848
π	= 3.141592654

TABLE 2.4
Miscellaneous

6,371,000 m	= approximate mean radius of earth
1.15 miles	= approximately 1 minute of latitude
69.1 miles	= approximately 1° of latitude
101 ft	= approximately 1 second of latitude
6 miles	= length and width of a normal township (U.S.)
36	= number of sections in a normal township (U.S.)
10,000 km	= distance from equator to pole (original basis for the length of the meter)

2.3.1 Metric Scales

The unit of measure usually used in metric scales is the millimeter, based on the International System of Units (SI).

2.3.2 Conversion Factors

Conversion factors for metric and imperial scales are listed in Tables 2.2–2.4).

2.3.3 Mapping Scales

Map scale is the ratio of "map distance" to ground distance. There are three methods of expressing scale:

1. *Numerical scale or representative fraction:* The numerical scale is the proportional length of a line on a map and the corresponding length on the earth's surface. This proportion is known as representative fraction (RF). The first number is a single unit of measure and the second number is the same distance on the ground (using the same units of measure).
   (RF) = distance on map/distance on ground
   Thus, if 1 inch on a map represents 1 mile on ground, the scale would be 1/63,360 or 1:63,360.
2. *Graphic scale:* The graphic scale is a geometric shape with divisions that represent increments of measure easily applied to the map. There are no standards for this type of scale.
3. *Verbal scale:* The verbal scale is usually expressed in a number of inches on the map equal to a number of feet on the ground.
   $1'' = 200'$

The term "small-scale map" indicates a large area of the earth shown in a map, typically requiring significant generalization of detail for map features (e.g., cities might be represented as point symbols). A map of the entire United States at a scale of 1:12,000,000 (where 1 inch equals 190 miles) is an example of a small-scale map. Conversely a "large-scale map" indicates a map that covers less geographic area and provides much greater map detail such as buildings and manholes. For example, a tax map of 1:2,400 (1 inch equals 200 ft) is a large-scale map.

2.4 ACCURACY VERSUS PRECISION

The terms *accuracy* and *precision* are frequently used in discussing standards and specifications. There is a recognized distinction between these two terms. Accuracy is the degree of closeness of an estimate to its true value, but unknown value and precision is the degree of closeness of observations to their means. Figure 2.1 illustrates various relationships between these two parameters. The true value is located at the intersection of the crosshairs, the center of the shaded area is the location of the mean estimate, and the radius of the shaded area is a measure of the uncertainty contained in the estimate.

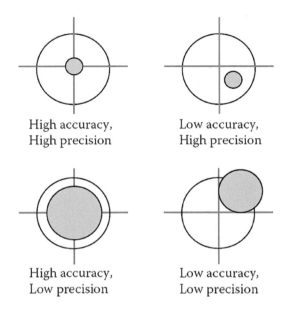

High accuracy,
High precision

Low accuracy,
High precision

High accuracy,
Low precision

Low accuracy,
Low precision

FIGURE 2.1   Accuracy versus precision.

2.4.1   GNSS ACCURACY MEASURES

When GNSS positions are logged over time, the positions are scattered over an area due to measurement errors. This dispersion of points is called a scatter plot, which GNSS manufacturers use to characterize their equipment's accuracy. The area within which the measurements or estimated parameters are likely to be is called the confidence region. The confidence region is then analyzed to quantify the GNSS performance statistically. The confidence region with a radius describes the probability that the solution will be within the specified accuracy. Two common GNSS accuracy measures are the *distance root mean square (DRMS)* and *circular error probability (CEP)*. These two measures are illustrated in Figure 2.2. Others are listed in Table 2.5.

DRMS is the radial or distance "root mean square" (RMS) error, calculated using the RMS values for the separate X and Y directions according to the formula:

$$\text{DRMS} = \sqrt{RMS_X^2 + RMS_Y^2}$$

The probabilities described by 1DRMS and 2DRMS are the typical 68.3% and 95% values, respectively. (2DRMS refers to twice the DRMS, irrespective of whether it is the 2D or 3D case; similarly, 3DRMS refers to thrice the DRMS.) Thus,

$$\text{DRMS} = \sqrt{\sigma_x^2 + \sigma_y^2}$$

$$2\text{DRMS} = 2\sqrt{\sigma_x^2 + \sigma_y^2}$$

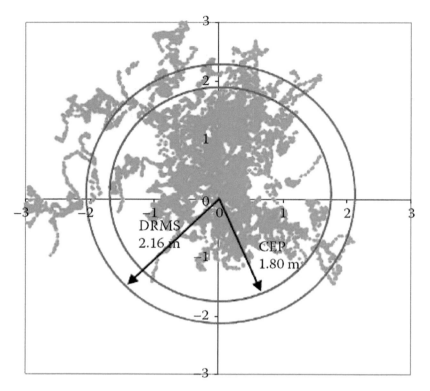

FIGURE 2.2   Position solutions from a NovAtel GPS receiver (courtesy: NovAtel Inc.). CEP refers to the radius of a circle in which 50% of the values occur (i.e., if a CEP of 5 m is specified, then 50% of horizontal point positions should be within 5 m of the true position). The term R95 is used for a CEP with the radius of the 95% probability circle. CEP $=0.62\sigma_x + 0.56\sigma_y$.

$$3\text{DRMS} = 3\sqrt{\sigma_x^2 + \sigma_y^2}$$

Another measure commonly used as an indicator of the likely quality of GNSS position results at a given location is the *dilution of precision (DOP)* factor (see an example computation in Section 6.4.4). DOP is a single value that provides a mathematical characterization of the user–satellite geometry at a specified location. It is related to the volume formed by the intersection of the user–satellite vectors, with the unit sphere centered on the user. Larger volumes give smaller DOP values, and vice versa.

Lower DOP values generally represent better position accuracy, but a lower DOP does not automatically mean a low position error. The quality of a GNSS position estimate depends upon both the measurement geometry as represented by DOP values, and range errors caused by signal strength, atmospheric effects, multipath, and so forth.

There are five types of DOP values that can be computed: geometric DOP, position DOP (PDOP), horizontal DOP (HDOP), vertical DOP (VDOP), and time DOP (TDOP).

TABLE 2.5

Selected Accuracy Measures

68.3	= percent of observations that are expected within the limits of one standard deviation (1-sigma[a])
1.65	= coefficient of standard deviation for 90% error
1.96	= coefficient of standard deviation for 95% error (2-sigma error)
CEPb	= GNSS circular error of probability (CEP) is determined by the number of points within a certain distance of a specific location as a percentage of the total number of points. So the CEP for 50% would be the distance within which half of the points would lie closer to a specific location (i.e., the value of the radius of a circle, centered at the actual position that contains 50% of the position estimates). Values stated as CEP apply to horizontal accuracy only.
RMS	= Root-mean-square (RMS) error is the value of one standard deviation (68%) of the error in one, two, or three dimensions.
PPMc	= parts per million (1/1 million = 0.000001). One PPM is 1 part in 1 million or the value is equivalent to the absolute fractional amount multiplied by 1 million.

[a] Example: <5 m 1-sigma.
[b] Example: 2.0 m CEP.
[c] Example 1: 2 cm + 1 PPM (× baseline length).
[c] Example 2: 5 mm + 0.5 PPM RMS.

The table in Figure 2.3 gives the multipliers required to go from one accuracy measure to another assuming that $\sigma_x/\sigma_y = 1$ and that VDOP/HDOP = 1.9 and PDOP/HDOP = 2.1.

2.4.2 EXAMPLES OF 2D ACCURACY MEASURES

Example 1: Given the Following Standard Deviations:

$\sigma_x = 1.3$ m
$\sigma_y = 1.5$ m

RMS (Vertical)	CEP	DRMS (Horizontal)	R95 (Horizontal)	2DRMS	RMS (3D)	
1	0.44	0.53	0.91	1.1	1.1	RMS (Vertical)
	1	1.2	2.1	2.4	2.5	CEP
		1	1.7	2.0	2.1	DRMS (Horizontal)
			1	1.2	1.2	R95 (Horizontal)
				1	1.1	2DRMS
					1	RMS (3D)

FIGURE 2.3 Equivalent accuracy multipliers.

The accuracy measures can be calculated as follows:

1. $CEP = 0.62 \times 1.3 + 0.56 \times 1.5 = 1.65\,m$
2. $DRMS = \sqrt{1.3^2 + 1.5^2} = 1.98\,m$
3. $2DRMS = 2\sqrt{1.3^2 + 1.5^2} = 3.96\,m$
4. $3DRMS = 3\sqrt{1.3^2 + 1.5^2} = 5.85\,m$

Example 2: How to Convert One Accuracy Measure to Another

When 1.8 m of CEP is given, what is the position accuracy for 2DRMS?
    Use the table in Figure 2.3 to find an equivalent multiplier for the conversion and follow the following steps.

1. Go down the "2DRMS" column to the "CEP" row.
2. The multiplier in this cell is 2.4.
3. $CEP = 1.8 \times 2.4 = 4.32\,m$.

Example 3: How to Interpret the Accuracy Measures

Which is more accurate, 1.8 m of CEP or 3 m of RMS (3D)?

1. In Figure 2.3, go down the "RMS (3D)" to the "CEP" row.
2. The multiplier in this cell is 2.5.
3. $RMS (3D) = CEP \times 2.5$.
4. $CEP = RMS (3D)/2.5 = 3/2.5 = 1.2$.

The position accuracy with 3 m of RMS (3D) will be 1.25 m of CEP accuracy. Therefore, 3 m of RMS (3D) is more accurate than 1.8 m of CEP.

2.4.3   3D Accuracy Measures

Similar to 2D accuracy measures there are many representations of 3D accuracy with various probabilities. 3D accuracy measures are conceptually similar to those in 2D expanded by 1D, the vertical accuracy. Spherical error probable corresponds to CEP in 2D, while mean radial spherical error corresponds to DRMS in 2D.

2.5   EQUIPMENT SPECIFICATIONS

Equipment manufacturers commonly list the precision and accuracy for conventional-angle– and distance-measuring devices and satellite GNSS positioning devices (Table 2.6). In the case of GPS, the best source for equipment specifications is the annual GPS receiver surveys by the *GPS World* magazine (www.gpsworld. com). For angle-turning instruments, the accuracy is normally reported as $\pm x$ number of seconds. For electronic distance-measuring devices (EDMs), the accuracy is defined as being $\pm x$ number of millimeters plus y parts per million (PPM). For example, a 1-km measurement made with a $\pm 5\,mm$ ± 5 PPM EDM will contain a $\pm 10\,mm$

TABLE 2.6
Typical Instrument Precisions

Type of Instrument	Precision (Typical)
Surveyor's compass	15 minutes
Builder's transit	1 minute
Mountain transit	30 seconds
Surveyor's theodolite	10 seconds
Control survey theodolite	1 second
Total station	5 seconds
Electronic distance meter (EDM)	5 mm (0.02 ft)
Precision level (optical/digital) standard deviation in height for 1 km two-way leveling	0.2 mm/1 km
Precise leveling rod[h]	0.7 mm (0.002 ft)/1 km[g]
Engineer's folding staff[h]	1.3 mm (0.004 ft)/1 km[g]
GPS (Autonomous[a])[f]	5–10 m
GPS (DGPS[b])[f]	0.5–2.5 m
GPS (RTK[c])[f]	1 cm + 1 PPM (horizontal) 2 cm + 1 PPM (vertical)
GPS (Kinematic[d])[f]	1 cm + 1 PPM (horizontal) 2 cm + 1 PPM (vertical)
GPS (Static[d])[f]	0.5–1 cm + 1 PPM (horizontal)[e] 0.5–2 cm + 1 PPM (vertical)[e]
GPS (Rapid-Static[d])[f]	1 cm + 1 PPM (horizontal) 2 cm + 1 PPM (vertical)

[a] Autonomous (code).
[b] Real-time differential (code).
[c] Real-time kinematic.
[d] Post-processed.
[e] Depends on observation length.
[f] Typical values obtained from 2010 GPS receiver survey (www.gpsworld.com).
[g] Values obtained from technical specifications by Trimble (www.trimble.com).
[h] Coded scale.

uncertainty. Likewise, the accuracy of levels is listed as being able to achieve a closure of x times the length of the level run. In most cases, GNSS precision is reported in similar expressions as the EDMs as well as in terms of accuracy measures that are specific to GNSS (e.g., RMS, CEP, and DRMS).

The equipment accuracy specifications supplied by manufacturers are a result of rigorous statistical testing involving thousands of measurements. In most surveying projects, surveyors can use theodolites having an accuracy specification of ±5 or 6 seconds and EDMs having an accuracy specification of ±5 mm + 5 PPM. More accurate instruments specified as "1 second" are available for high precision work. Real-time kinematic (RTK) GNSS with accuracy specifications of ±1 cm (horizontal) and ±2 cm (vertical) are sufficient for detail topographic (engineering type) surveys. For geodetic control surveys,

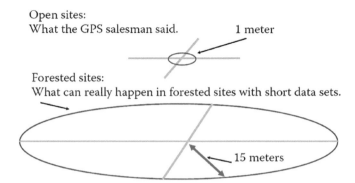

FIGURE 2.4 Interpretation of manufacturer-specified GNSS precision. GNSS manufacturer specifications are for open sites and long data sets.

static survey receivers with specifications of ±0.1–0.5 cm (horizontal) are sufficient— the actual accuracy depends on the site characteristics and data length (Figure 2.4).

Digital self-leveling levels with automated data reduction and adjustment software that are capable of achieving a closure of ±0.01 ft in 1 mile of double run levels are commonly used by surveyors.

The manufacturer's accuracy specification refers to the plus-or-minus uncertainty in each *direction* or *pointing* (not in each angle), a direction being composed of a direct and reverse pointing to a single target. An angle is then considered to be the difference between two directions or pointings.

2.6 LIMITS OF CLOSURE

Limits of closure (also known as *precision ratios*) is a method for evaluating survey accuracy that is well known and published in many textbooks. Traditionally, the allowable limits of closure for surveys of U.S. federal lands are derived from the summation of all of the latitudes and departures along the surveyed lines of a closed traverse. Various manuals and standards have, in the past, defined survey accuracy in terms best expressed by precision ratios (see, e.g., Appendix B). However, since the advent of modern computers, traverses are computed using coordinates, as opposed to latitudes and departures. The reciprocal of the amount of linear misclosure divided by the length of the traverse is called the *error of closure* and is expressed as a ratio. For example, if the linear misclosure of a traverse is 0.484 ft and the length of the traverse is 9,551.45 ft, the error of closure will be expressed as 1/0.00005, or approximately 1:19,740. This is commonly stated as 1 in 19,740 or 1 part in 19,740 parts.

The error of closure (or loop closure) concept is also applied to GNSS networks. In such cases, loop closure is a procedure by which the internal consistency of a GNSS network is assessed. A series of baseline components from more than one GNSS session, forming a loop or closed figure, are added together. The *closure error* is then the ratio of the length of the line representing the combined errors of all the vectors' components to the length of the perimeter of the network figure. Any loop closures that only use baselines derived from a single common GNSS *session* will yield an apparent error of zero, because they are derived from the same simultaneous GNSS observations.

If a project requires the publication of misclosure as part of a survey, the computation can be done in a traditional sense for a quick check. The project data can then be recomputed using mathematically rigorous least squares.

2.7   LEAST SQUARES ANALYSIS

The least squares method of analysis of survey measurements is now commonly used in all aspects of surveying. Every surveyor who surveys the boundaries of a property has computer hardware and software available to perform least squares analysis and adjustment of survey data. RTK GNSS makes use of this process in the field to resolve baseline measurements on the fly. RMS error is evaluated in the RTK GNSS survey data logging device in the field. Statistical methods of data analysis are also used in many other geomatics-related professions. When the various data from different sources are combined in a GIS, one of the first questions that comes to mind is how accurate are the data. How closely does the virtual picture of reality mimic the real world or actual conditions in the field?

All measurements are prone to random errors, systematic errors, and mistakes (blunders). Past survey guidelines expressed survey quality standards in the form of a closure (precision) ratio. Using a precision ratio to evaluate survey error has a well-defined place in determining the relative precision of past surveys. It is a well-understood principle that during the course of a dependent resurvey the limit of closure or standard in place at the time of the original survey is how past survey measurements are judged today. It is because of this that surveyors need to continue to evaluate resurvey data and calculate precision ratios or loop closures for their work. However, use of precision ratios has their shortcomings:

> This method of quantifying error makes no attempt to identify measurement mistakes, or impart any information as to the positional error associated with any particular corner point of a survey or dependent resurvey. Precision ratios serve only to imply the general quality of the relative precision of a closed traverse. The loop closure has minimum redundancy and does not evaluate scale or rotational errors.
>
> *(Craig and Wahl, 2003, 92)*

Numerous general methods are available to disclose error in survey measurements. For instance, three angles measured in a plane triangle must equal 180°. The sum of the angles measured around the horizon at any point must equal 360°, and the sum of latitudes and departures must equal zero for closed traverses that begin and end at the same point. Each of these conditions involves one redundant measurement. In the case of three angles of a plane triangle, if only two angles were measured, angles A and B, the third angle, C, could be computed as $C = 180° - A - B$. The actual measurement of the angle is redundant but allows the surveyor to assess the errors in the measurements made. The total angular error could be distributed by adjusting the angles and forcing the sum of the angles of the triangle to equal 180°. This adjustment of the measured data would result in statistically improved precision. There are many different ways to adjust survey measurement data; some are more arbitrary than others.

In surveying, redundant measurements are very important. Prudent surveyors check the magnitude of the error of their work by making redundant measurements. These extra measurements allow the surveyor to assess errors and accept or reject measurements. They also make valid adjustment of survey measurements possible. The more a measurement is validated by additional measurements, the greater is the likelihood of the measurement approaching a true value. While the process of adjusting a plane triangle is relatively simple, the process becomes much more complex when analyzing large survey networks. Adjustments correct measured values, so they are consistent throughout the network. Many methods for adjusting data have been developed, but the least squares method has significant advantages over all of them.

Least squares adjustment is based on the mathematical theory of probability and the condition that *the sum of the squares of the errors times their respective weights is minimized*. The least squares adjustment is the most rigorous of adjustments yet, it is applied with greater ease than other adjustments because it is not biased. Least squares enable rigorous post-adjustment analysis of survey data and can be used to perform presurvey planning (see, e.g., Section 6.4.4).

The most important aspect of using least squares is that surveyors can analyze all types of measurements simultaneously. This could include horizontal and slope distances, vertical and horizontal angles, azimuths, vertical and horizontal control coordinates, and GNSS baseline observations. Least squares adjustments also allow for the application of "relative weights" to properly reflect the expected reliability of different measurement types. An example would be weighting a line measured with a tape differently than one measured with GNSS.

> Least squares analysis has the advantage that after an adjustment has been finished, a complete statistical analysis can be made from the results. Based on the sizes and distribution errors, various tests can be conducted to determine if a survey meets acceptable tolerances or whether measurements must be repeated. If blunders exist in the data, these can be detected and eliminated. Least squares analysis enables precisions for the adjusted quantities to be determined easily, and these precisions can be expressed in terms of error ellipses for clear and lucid depiction.
>
> *(Wolf and Ghilani, 1997, Section 1.7, 9)*

When computing loop closures of a closed traverse, precision ratios can only imply the general magnitude of the error. Using least squares adjustments, surveyors can express error in terms of positional tolerance of a single point, the relative error of all of the points in a network, or the range of precision within a large network.

2.8 MAPPING AND GIS STANDARDS

2.8.1 MAP SCALE

Map scale specifies the amount of reduction between the real world and the graphic representation on a map. It is usually expressed graphically, as a fraction (1/20,000), a ratio (1:20,000), or equivalence (1 mm = 20 m). Because map scale is most often used to describe paper map products, it is often assumed that the scale is fixed and cannot change. However, a digital map in a GIS can be reduced or enlarged on the screen by zooming in or out. This implies that geographic data in a GIS does not really have a true "map scale."

When scale is used to describe digital data, it is often referring to the scale of the source data or the scale at which the digital data looks "right." As a result, this display scale influences the amount of detail that can be shown. Digital data viewed at inappropriate display scales within a GIS can be misleading. A map or view can be created in the GIS that have a scale well beyond the accuracy of the original mapping, thus misrepresenting the accuracy of spatial relationships between objects.

Map scale is defined by the U.S. Geological Survey (USGS Fact Sheet 038-00, April 2000) as follows:

> To be most useful, a map must show locations and distances accurately on a sheet of paper of convenient size. This means that all things included in the map—ground area, rivers, lakes, roads, distances between features, and so on—must be shown proportionately smaller than they really are. The proportion chosen for a particular map is its scale.

2.8.1.1   How to Interpret Numerical Map Scales

When thinking of larger or smaller scale, it is better to think of the map scale as a fraction. As the number represented by the fraction gets larger, so does the scale. Conversely, as the denominator of the fraction gets larger, the scale gets smaller. A map at a scale of 1:100,000 ($1'' = 8,333'$) is a smaller scale map than a map at a scale of 1:24,000 ($1'' = 2,000'$). Similarly, a map at a scale of 1:2,400 ($1'' = 200'$) is a smaller scale map than a map at a scale of 1:1,200 ($1'' = 100'$).

Today many maps are created in a computer environment where maps can be plotted at virtually any scale of choice. Scale remains an important factor in the accuracy of a map. Many digital maps are derived from aerial photography or digitized from existing paper maps. In such instances, the map accuracy is a function of the scale of the original aerial photography or map. However, in cases where some features in a digital map have been located using very accurate GNSS surveys, the accuracy of the GNSS-surveyed features is largely unrelated to the scale of the map.

2.8.2   Map Resolution

Map resolution refers to the accuracy of the location and shape of a map feature shown at a given scale. In general, as map scale increases (e.g., 1:100 k to 1:50 k to 1:20 k), so do map resolution and accuracy. However, accuracy is also affected by the quality of source data used to map a feature.

Features on large-scale maps more closely represent the real world because the amount of reduction (from real world to map) is less. As the level of detail of a paper map increases for a given area of earth, the size of the paper map required to cover the same area also increases. Similarly, as digital map resolutions become more detailed and accurate, file sizes increase because more information is now represented for the same area.

Spatial data can never be any more accurate than the original source from which the data were acquired (and frequently it is less accurate, depending on the method of data conversion). Therefore, if data were digitized from a source map scale of $1'' = 2,000'$, and a map was created at $1'' = 100'$, the map accuracy of features shown in the map is still $1'' = 2,000'$.

2.8.3   MAP ACCURACY

Map accuracy should be determined by the intended use of the map. Historically, map accuracy determined the scale at which the map would be drawn. Until recently, it has been customary to specify the scale of aerial photography for digital orthophotos, planimetric features, and topographic features, and then apply the National Map Accuracy Standards (NMAS) or other similar standard to determine the accuracy of the map.

Recent trends, however, are to treat accuracy as a property of the map to be reported, rather than a specification for producing the map. FGDC-STD-007-1998, Geospatial Positioning Accuracy Standards, Part 3: National Standard for Spatial Data Accuracy specifies testing methodology and reporting requirements for map accuracy. Section 3.1.4 of this standard states:

> Data producers may elect to use conformance levels or accuracy thresholds in standards such as the National Map Accuracy Standards of 1947 (U.S. Bureau of the Budget, 1947) or Accuracy Standards for Large-Scale Maps [American Society for Photogrammetry and Remote Sensing (AS-PRS) Specifications and Standards Committee, 1990] if they decide that these values are truly applicable for digital geospatial data.

The horizontal accuracy of a map is related to the map scale. According to the U.S. NMAS (issued by the U.S. Bureau of the Budget June 10, 1941, and revised June 17, 1947), the horizontal accuracy of a map is defined by the following specifications:

> For maps on publication scales larger than 1:20,000, not more that 10 percent of the points tested shall be in error by more than 1/30 inch, measured on the publication scale; for maps on publication scales of 1:20,000 or smaller, 1/50 inch.

Vertical accuracy of contour mapping is related to the contour interval, not map scale. The same publication defines vertical accuracy for contour maps as:

> not more than 10 percent of the elevations tested shall be in error more than one-half the contour interval.

Accuracy relates to how well features on the map correspond to their "real-world" counterpart. Table 2.7 lists spatial accuracy values associated with various mapping scales, based on the NMAS.

Tables 2.8 and 2.9, taken from the U.S. Army Corps of Engineers Topographic Accuracy Standards (EM 1110-1-1005), detail American Society for Photogrammetry and Remote Sensing (ASPRS) horizontal and vertical accuracy requirements.

TABLE 2.7

Map Accuracies Based on U.S. NMAS

Scale	No. of Feet to the Inch	Horizontal Accuracy (ft)	Vertical Accuracy (ft)
1:1,200	100	3.33	[Contour Interval]×0.5
1:2,400	200	6.67	[Contour Interval]×0.5
1:4,800	400	13.33	[Contour Interval]×0.5
1:12,000	1000	33.33	[Contour Interval]×0.5
1:24,000	2000	40	[Contour Interval]×0.5
1:100,000	8333	166.67	[Contour Interval]×0.5

TABLE 2.8

ASPRS Planimetric Feature Coordinate Accuracy Requirement for Well-Defined Points

Target Map Scale		Limiting RMS Error in X or Y, ft ASPRS		
No. of Feet to the Inch	Ratio ft/ft	Class 1	Class 2	Class 3
5	1:60	0.05	0.10	0.15
10	1:120	0.10	0.20	0.30
20	1:240	0.20	0.40	0.60
40	1:480	0.40	0.80	1.20
60	1:720	0.60	1.20	1.80
100	1:1,200	1.00	2.00	3.00
200	1:2,400	2.00	4.00	6.00
400	1:4,800	4.00	8.00	12.00
800	1:9,600	8.00	16.00	24.00
1,000	1:12,000	10.00	20.00	30.00

TABLE 2.9

ASPRS Topographic Elevation Accuracy Requirement for Well-Defined Points

Target C.I.[a] (ft)	ASPRS Limiting RMS Error, ft					
	Topo Feature Points			Spot or DTM Elevation Points		
	Class 1	Class 2	Class 3	Class 1	Class 2	Class 3
0.5	0.17	0.33	0.50	0.08	0.16	0.25
1	0.33	0.66	1.0	0.17	0.33	0.5
2	0.67	1.33	2.0	0.33	0.67	1.0
4	1.33	2.67	4.0	0.67	1.33	2.0
5	1.67	3.33	5.0	0.83	1.67	2.5

[a] Contour interval.

The NMAS and ASPRS require different accuracy standards for the same scale. In addition, the ASPRS places multiple accuracy standards for the same scale depending on which class is chosen. For example, at a scale of 1:2,400, NMAS reports 6.67 ft horizontal accuracy and 1 foot vertical accuracy (half of the contour interval). For the same scale, the ASPRS Class 2 reports 4 ft horizontal accuracy and 1.33 ft vertical accuracy for topographic features.

2.9 CLASSICAL SURVEYING STANDARDS

Standards are used to evaluate and judge the quality of survey work. They are statements of required accuracies of survey measurements, such as traverse closure for horizontal measurements and level loop closure for vertical measurements. Specifications, on the other hand, will detail the necessary steps (or procedures) to achieve a particular standard. In the United States, the National Geodetic Survey (NGS) has implemented the most widely accepted set of standards and specifications for survey classifications.

The *Standards and Specifications for Geodetic Control Networks* (see partial details in Appendix B) initially issued in September 1984 by the Federal Geodetic Control Subcommittee (FGCS) includes the familiar first-, second-, and third-order classifications of conventional surveys. These classifications use precision ratios—the older form of accuracy expression (see, e.g., Section 2.6). For instance, the ratio 1:10,000 means that the error tolerance will be no greater than 1 unit for every 10,000 units of horizontal measurement.

Although the FGCC standards and specifications were developed primarily to support geodetic control surveys, third-order requirements were designed for local mapping and engineering projects. For the land surveyor using conventional equipment, the specifications for angular, linear, and vertical closure are of great importance. However, due to satellite GNSS and widespread availability of least squares adjustment software, it has been necessary to classify surveys by positional tolerance (Section 2.10). The following paragraphs summarize some selected specifications for angular, linear, and vertical closures for conventional surveying as stated in the FGCC publication (Appendix B).

Horizontal first- and second-order specifications. These were developed for most accurate work. The triangulation network that covered the United States prior to GPS was composed primarily of first- and second-order stations. In a conventionally observed triangulation network, the maximum triangle closure for first order shall not exceed 3″. For traverse networks, the maximum angular closure for first order shall be $1.7''\sqrt{N}$, where N is the number of legs in a traverse. The minimum position closure in a first-order traverse shall be no less than 1:100,000.

Horizontal third-order specifications. In triangulation, the maximum triangle closure for third-order class II shall not exceed 10″. For traverse networks, the maximum angular closure for third-order class I shall be $10''\sqrt{N}$ and class II shall be $10''\sqrt{N}$, where N is the number of legs in a traverse. Minimum position closures in a traverse are to be no less than 1:10,000 for class I and no less than 1:5,000 for class II.

Vertical first- and second-order specifications. For level loop closures the FGCC publication states a first-order accuracy requirement of $4\,\text{mm}\,\sqrt{F}$ and a second-order class II requirement of $6\,\text{mm}\,\sqrt{F}$, where F is the length of the level loop in kilometers.

Vertical third-order specifications. The third-order class II accuracy requirement for level loop closures is $12\,\text{mm}\,\sqrt{F}$, where F is the length of the level loop in kilometers. This requirement can also be stated as $0.05\,\text{ft}\,\sqrt{M}$, where M is the length of the level loop in miles.

In the vertical accuracy specifications, the maximum allowable loop misclosures are for the assessment of results in differential leveling prior to any least squares adjustments. In addition to these standards, Table 2.10 specifies the maximum relative elevation errors allowable between two control points (or benchmarks) as determined by a weighted least-squares adjustment. As an example, elevations for two control points 25 km apart, established by second-order class I standards, should be correct to within $\pm 1.0\sqrt{25} = \pm 5$ mn.

2.10 GPS SURVEYING STANDARDS

In many ways, GPS is nowadays the better method of choice for observing or establishing horizontal control networks and, to a lesser extent, vertical networks as well. For instance, it is no longer necessary to observe a triangulation network using classical optical surveying methods. In fact, it would be less productive to do so if you have GPS at your disposal. By 1985, it was possible to achieve a 1,000 times better accuracy than those specified in the old first-order without a corresponding 1,000-fold increase in equipment, training, personnel, or effort. Relative accuracies exceeding 1:100,000 could be easily obtained from just a few minutes of GPS observation. For this reason, the FGCS developed new classifications to include GPS surveys (Table 2.11). The new categories for GPS surveys included AA, A, and B order, with corresponding relative accuracies of 1:100,000,000, 1:10,000,000, and 1:1,000,000. A lower order of accuracy for GPS surveys, identified as Order C, overlaps three orders of accuracy applied to traditional horizontal surveys.

To meet various local needs for surveyors, engineers, and scientists, governments and agencies have established national reference networks consisting of control monuments and benchmarks. In the United States, a National Spatial Reference System (NSRS)

TABLE 2.10
Vertical Control Survey Accuracy Standards (FGCC 1984)

Order and Class	Relative Accuracy Required between Benchmarks
First Order	
Class I	$0.5\,\text{mm} \times \sqrt{K}$
Class II	$0.7\,\text{mm} \times \sqrt{K}$
Second Order	
Class I	$1.0\,\text{mm} \times \sqrt{K}$
Class II	$1.3\,\text{mm} \times \sqrt{K}$
Third order	$2.0\,\text{mm} \times \sqrt{K}$

Note: K is the distance between benchmarks in kilometers.

TABLE 2.11

Horizontal Control Survey Accuracy Standards (FGCS 1984 and 1985)

GPS Order[a]	Traditional Surveys[b] Order and Class	Relative Accuracy between Points
Order AA		1:100,000,000
Order A		1:10,000,000
Order B		1:1,000,000
Order C-1	First order	1:100,000
	Second order	
Order C-2-I	Class I	1:50,000
Order C-2-II	Class II	1:20,000
	Third order	
Order C-3	Class I	1:10,000
	Class II	1:5,000

[a] Published in 1985.
[b] Published in 1984.

consists of more than 270,000 horizontal control monuments and approximately 600,000 benchmarks nationwide. The primary uses of horizontal control are as follows:

1. GPS-surveyed control points that meet the Order AA and Order A standards are common in global, national, and regional networks that are primarily used for geodynamic and deformation studies.
2. GPS-surveyed points that densify the network within areas surrounded by primary control are executed to GPS Order B standards. These networks are common in high-value land areas and are commonly used for high precision engineering surveys.
3. Survey control to meet mapping, GIS, property surveys, and engineering needs are set by traverse and triangulation to first- and second-order stations, and by GPS to Order C standards.
4. Control for local construction projects and small-scale topographic mapping are referenced to higher order control monuments and, depending on accuracy requirements, may be set to third-order class I or third-order class II standards.

Despite the advantages of GPS surveying, larger errors in the vertical component make it less favorable for heighting in comparison to classical methods. GPS is less accurate in the vertical direction due to compounding effects of the different layers of the atmosphere on the GPS signals as they travel from the satellite to the receiver. Furthermore, GPS measures heights above a global mathematical surface called the *ellipsoid*, and thus the need for a reliable geoid model in order to convert GPS heights into orthometric heights. Through long-term observations (e.g., lasting several hours), it is possible to average out the systematic errors. However, spirit leveling is still the most accurate method to transfer orthometric heights.

Craig and Wahl (2003) have identified three very important requirements of a new standard, that is, it should be *technology-neutral, inclusive*, and *understandable*.

In other words, an ideal standard should be developed with the idea that it can be applied to both old and new technology, it should not exclude major technologies that are currently considered acceptable, and it should be usable and understandable rather than confusing or ambiguous. The 1998 FGCS standards for control points as listed in Table 2.12 seem to address the three requirements reasonably well. These new standards are independent of the method of survey and are based on a 95% confidence level. In order to meet these standards, control points in the survey must be consistent with all other points in the network. For horizontal surveys, the accuracy standards specify the radius of a circle within which the true or theoretical location of the survey point falls 95% of the time. For leveling, the vertical accuracy standards specify a linear value (plus or minus) within which the true or theoretical location of the point falls 95% of the time. Procedures leading to classification according to these standards involve four steps (Ghilani and Wolf, 2007, 539–540):

1. The survey observations, field records, sketches, and other documentation are examined to ensure their compliance with specifications for the intended accuracy of the survey.
2. A minimally constrained least-squares adjustment of the survey observations is analyzed to guarantee that the observations are free from blunders and have been correctly weighted.
3. The accuracy of control points in the local existing network to which the survey is tied is computed by random error propagation and weighted accordingly in the least-squares adjustment of the survey network.
4. The survey accuracy is checked at 95% confidence level by comparing minimally constrained adjustment against established control.

TABLE 2.12

Accuracy Standards: Horizontal, Ellipsoid Height, and Orthometric Height (FGCS 1998)

Accuracy Classifications	95% Confidence Less Than or Equal To (m)
1 mm	0.001
2 mm	0.002
5 mm	0.005
1 cm	0.010
2 cm	0.020
5 cm	0.050
1 dm	0.100
2 dm	0.200
5 dm	0.500
1 m	1.000
2 m	2.000
5 m	5.000
10 m	10.000

2.11   OTHER STANDARDS

Table 2.13 summarizes the various standards for surveying, mapping, GIS, and remote sensing. Although far from being a complete list of existing standards, this table is fairly representative of the historical developments. Take, for instance, a portion of the standards developed by the Federal Geographic Data Committee (FGDC) that apply to control surveys. The draft FGDC Geodetic Subcommittee standard is an effort to improve the older FGCC 1984 standards and the 1989 FGDC *Geodetic Geometric Accuracy Standards and Specifications for Using GPS Relative Positioning Techniques*. It describes a general scheme of classification that is based on reporting coordinate data, with associated positional tolerances, specifically the

TABLE 2.13
Accuracy Standards and Specifications

U.S. National Control Survey Accuracy Standards
1. Federal Geographic Data Committee (FGDC) Geometric Geodetic Accuracy Standards and Specifications for Using GPS Relative Positioning Techniques (Version 5.0, 1989)
2. FGDC Geospatial Positioning Accuracy Standards
   a. FGDC-STD-007.1-1998 Part 1: Reporting Methodology
   b. FGDC-STD-007.2-1998 Part 2: Standards for Geodetic Networks
   c. FGDC-STD-007.3-1998 Part 3: National Standards for Spatial Data Accuracy
   d. FGDC-STD-007.4-2002 Part 4: Architecture, Engineering, Construction, and Facilities Management
   e. FGDC-STD-007.5-2005 Part 5: Standards for Nautical Charting Hydrographic Surveys
   f. FGDC-STD-008-1999 Content Standard for Digital Orthoimagery
   g. FGDC-STD-009-1990 Content Standard for Remote Sensing Swath Data
   h. FGDC-STD-010-2000 Utilities Data Content Standard
   i. FGDC-STD-011-2001 Standard for U.S. National Grid
   j. FGDC-STD-012-2002 Content Standard for Digital Geospatial Metadata: Extensions for Remote Sensing Metadata
U.S. National Mapping Accuracy Standards
1. National Map Accuracy Standards (NMAS 1947)
2. National Standard for Spatial Data Accuracy (NSSDA)
3. American Society for Photogrammetry and Remote Sensing (ASPRS) Standards for Large Scale Maps (ASPRS 1990)
4. ASPRS LIDAR Guidelines Vertical Accuracy Reporting for LIDAR Data
U.S. Land Title National Survey Accuracy Standards
1. American Land Title Association (ALTA/ACSM) ALTA Accuracy Standards
2. U.S. Forest Service and U.S. Bureau of Land Management Standards and Guidelines for Cadastral Surveys Using Global Positioning System Methods (version 1.0, May 2001)
U.S. State Highway Control Survey Accuracy Standards
1. Various U.S. State Highway Department Survey Standards
   a. GPS 3D Survey Standards
   b. Conventional Survey Standards
International Survey and Mapping Accuracy Standards
1. Australian Inter-Governmental Committee on Surveying and Mapping Standards and Practices for Control Surveys (SP1 version 1.5, May 2002)

relative error circle reported at 95% confidence. Two sets of values to be reported are described as network accuracy and local accuracy. These two values are defined as follows:

> *Network accuracy* of a control point is a value that represents the uncertainty in the coordinates of the control point with respect to the geodetic datum at the 95% confidence level. For NSRS network accuracy classification, the datum is considered to be best expressed by the geodetic values at the Continuously Operating Reference Stations (CORS) supported by NGS. By this definition, the local and network accuracy values at CORS sites are considered to be infinitesimal, that is, to approach zero.
>
> *Local accuracy* of a control point is a value that represents the uncertainty in the coordinates of the control point relative to the coordinates of other directly connected, adjacent control points at the 95% confidence level. The reported local accuracy is an approximate average of the individual local accuracy values between this control point and other observed control points used to establish the coordinates of the control point.

Historically, there have been numerous attempts to create new standards to reflect both current accuracy needs and new technology. Various approaches have been tried, including loop closures, theoretical uncertainty, positional tolerance, and other mixed standards, which have evolved toward the FGDC type of standard. One example is the standard published by the American Land Title Association (ALTA; http://www.acsm.net/alta.html), which follows an error propagation type model. The 1999 ALTA standard defines positional uncertainty and positional tolerance:

> *Positional uncertainty* is the uncertainty in location, due to random errors in measurement, of any physical point on a property survey, based on the 95% confidence level.
>
> *Positional tolerance* is the maximum acceptable amount of positional uncertainty for any physical point on a property survey relative to any other physical point on the survey, including lead-in courses.

Internationally, efforts such as those by the Australian Inter-Governmental Committee on Surveying and Mapping have led to other national standards for control surveys.

BIBLIOGRAPHY

Craig, B. A. and J. L. Wahl. 2003. Cadastral survey accuracy standards. Surveying and Land Information Science (SaLIS), 63 (2): 87–106.

Dewberry, S. O. 2008. Land Development Handbook: Planning, Engineering, and Surveying. New York: McGraw-Hill.

Genovese I. (ed.). 2005. *Definitions of Surveying and Associated Terms*. Gaithersburg, MD: American Congress on Surveying and Mapping.

Ghilani C. D. and P. R. Wolf. 2007. Elementary Surveying: An Introduction to Geomatics. Upper Saddle River, NJ: Prentice Hall.

GPS World. *GPS Receiver Survey 2010*. GPS World http://www.gpsworld.com, (accessed on January 26, 2010).

Hofmann-Wellenhof, B., H. Lichtenegger and J. Collins 1994. *GPS Theory and Practice*. 3rd ed. New York: Springer-Verlag.

McKay, E. J. (ed.). 2001. *NGS Positioning Accuracy Standards*. Silver Spring, MD: National Geodetic Survey.

PaMagic. 2002. *Local Government Handbook for GIS*. Pennsylvania: Commonwealth of Pennsylvania.

U.S. Army Corps of Engineers. Chapter 2, topographic accuracy standards, EM 1110-1-1005. August 3, 1994.

Van Sickle, J. 2007. *GPS for Land Surveyors*. 3rd ed. Boca Raton, FL: CRC Press.

Wolf, P. R. and C. D. Ghilani. 1997. Adjustment Computations: Statistics and Least Squares in Surveying and GIS. New York: Wiley.

EXERCISES

1. The following is an example of horizontal accuracy standards in the 1984 Standards and Specifications for Geodetic Control Networks by the Federal Geodetic Control Committee:

Order of Accuracy	Maximum Closure
First	1:100,000
Second-class I	1:50,000
Second-class II	1:20,000
Third-class I	1:10,000
Third-class II	1:5,000

   What is the first-order horizontal accuracy of relationship between two stations that are 13,786 m apart?
   a. 100,000/13,786 = 7.254 m
   b. 13,786/100,000 = 0.138 m
   c. 13,786/100,000 = 0.138 mm
   d. 100,000/13,786 = 7.254 mm

2. Complete the following table and classify each line according to FGCC 1984 order and class. In the table s represents propagated standard deviation of distance between survey points obtained from a minimally constrained least-squares adjustment, and d is the distance between survey points.

Line	S (mm)	d (m)	$b(mm)/\sqrt{(km)}$	Class-Order
1-2	1.574	1,718	....	....
1-3	1.743	2,321	....	....
2-3	2.647	4,039	....	....
3-4	2.127	3,039	....	....
4-1	5.125	1,819	....	....

3. Complete the following table and classify each line according to FGCC 1984 order and class. In the table d represents the approximate horizontal distance between control point positions traced along existing level routes, and S is the propagated standard deviation of elevation difference in millimeters between survey control points obtained from a minimally constrained least-squares adjustment.

Line	s (m)	d (miles)	1:a	Order-Class
1-2	0.141	10.6	....	....
1-3	0.170	12.5	....	....
2-3	0.164	9.6	....	....
3-4	0.235	15.4	....	....
4-1	0.114	10.1	....	....

4. The output map scale is an important specification when designing a mapping project. The output map scale greatly affects project costs and scheduling. Which of the following is a correct interpretation?
   a. The larger the scale map required (i.e., $1'' = 100'$ is smaller than $1'' = 200'$) by a client, the shorter it will take to produce and the less costly it will be.
   b. The larger the scale map required (i.e., $1'' = 100'$ is larger than $1'' = 200'$) by a client, the shorter it will take to produce and the less costly it will be.
   c. The smaller the scale map required (i.e., $1'' = 100'$ is smaller than $1'' = 200'$) by a client, the longer it will take to produce and the less costly it will be.
   d. The larger the scale map required (i.e., $1'' = 100'$ is larger than $1'' = 200'$) by a client, the longer it will take to produce and the more costly it will be.

5. The product accuracy significantly increases project costs and schedules. Mapping clients often fall short of meeting their goals by assigning the wrong product accuracy specifications to their project. Select the correct statement.
   a. Assigning a very strict product accuracy (e.g., ASPRS Class I) decreases the amount of mapping data.
   b. Assigning a very strict product accuracy (e.g., ASPRS Class I) lowers the mapping scale.
   c. Assigning a very strict product accuracy (e.g., ASPRS Class I) implies larger contour interval.
   d. Assigning a very strict product accuracy (e.g., ASPRS Class I) increases the amount of mapping data.

6. Choosing the correct photo scale, which is the flying height (above ground level) divided by the camera's focal length, is the key to a successful aerial mapping project. Which of the following is incorrect:
   a. The required contour interval will affect the photo scale.
   b. A lower photo scale will result in an increased cost for aerial photography acquisition.
   c. The project cost would be reduced significantly by eliminating the contour component.
   d. A lower photo scale will result in a reduced cost for aerial photography acquisition.

7. _____ accuracy of a control point represents the uncertainty in the coordinates of the control point relative to the coordinates of other directly connected, adjacent control points at the 95% confidence level.
   a. Local
   b. Network
   c. Global
   d. Standard

8. In this question and the next, you need to read the article quoted before you can be able to answer the question. One of the three requirements of a cadastral standard is that it should be technology-neutral (Craig and Wahl, 2003). The other two requirements are that:
   a. A standard should be *inclusive* and *understandable*.
   b. A standard should be *technology-specific* and *understandable*.
   c. A standard should be *technology-specific* and *usable*.
   d. A standard should be *technology-specific* and *exclusive*.

9. Which of the following is not true about the application of a standard (Craig and Wahl, 2003)?
   a. A standard can be applied as a *design tool* or *variable*.
   b. A standard can be applied as a *systematic error*.
   c. A standard can be applied as an *evaluation tool*.
   d. A standard can be applied as a *project requirement*.

10. Given the standard deviations:

    $\sigma_x = 1.7\,cm$

    $\sigma_y = 2.5\,cm$

    Compute the following GNSS accuracy measures in meters: RMS, CEP, R95, and 2DRMS. What would be the accuracy classification based on Table 2.12?

3 Professional and Ethical Responsibilities

Know, first, who you are...

—Epictetus

3.1 KNOW WHAT YOU DO

Being in a profession provides you with certain rights and privileges that belong only to the members of the profession.

A surveyor is defined by International Federation of Surveyors (FIG) as

> a professional person with the academic qualifications and technical expertise to practise the science of measurement; to assemble and assess land and geographic related information; to use that information for the purpose of planning and implementing the efficient administration of the land, the sea and structures thereon; and to instigate the advancement and development of such practises.

The term *geomatics* (or *geomatics engineering*) is now more commonly used than *land surveying* in the industry. There are many credible sources for the definitions of the term, including professional organizations and the Web sites of university departments.

Geomatics engineers (surveyors) measure and collect data on specific areas of land. Once the data is interpreted, it is used for a variety of purposes. The information and analysis have a key role in a diverse range of sectors, including construction, property, cartography, engineering, geosciences, exploration, and geographic information systems.

3.1.1 TYPICAL WORK ACTIVITIES

Traditionally, geomatics engineers (surveyors) measure and record physical characteristics of land, such as contour and elevation, and determine property boundaries by researching legal documents and performing on-site measurement. Working for construction companies, engineering firms, and government agencies, they assist in planning land development and real estate sales. In addition to assessing land for development, geomatics engineers survey a range of different areas and infrastructure such as bridges, dams, highways, airports, landfill sites, pipeline and distribution systems, sports complexes, wetlands, and waterways. Typical work activities include

- Discussing specific project requirements with clients;
- Measuring the ground as required by the client (including aspects such as small- and large-scale distances, angles, elevations, etc.);

- Gathering data on the earth's physical and man-made features through surveys;
- Processing data;
- Undertaking digital mapping;
- Producing detailed information (subsequently analyzed by planners, builders, and cartographers);
- Using a range of equipment to produce surveys, including Global Positioning System and conventional methods;
- Analyzing information thoroughly before it is handed over to other professionals;
- Thinking creatively to resolve practical planning and development problems;
- Interpreting data using maps, charts, and plans;
- Utilizing data from a range of sources, such as aerial photography, satellite surveys, and laser beam measuring systems;
- Using computer-aided design (CAD) and other IT software to interpret data and present information;
- Keeping up to date with new and emerging technology; and
- Providing advice to a range of clients.

Licensed geomatics engineers/professionals are involved in the managing and monitoring of projects from start to finish.

3.1.2 WORK CONDITIONS

Geomatics work typically involves both field and office components, and traveling time is often paid for work away from home. Working hours are typically nine to five, but this can vary depending on location and whether the work is based locally, nationally, or overseas. Weekend or shift work is sometimes required, and work may take place in all weather conditions. It is possible to become self-employed, although salaried employment is preferred by majority as it is not always easy to find work on one's own. Usually work must be completed to deadlines, which can, on occasion, be stressful. In 2008, the typical nongraduate starting salary was about $18 per hour, or $38,000 annually, whereas a graduate could expect to receive a starting salary of about $27 per hour, or $56,000 annually (U.S. Bureau of Labor Statistics, 2010, 160). For licensed surveyors, the typical starting salary range was likely to be $70,000–$80,000, while the range of typical salaries at senior level, including management/partnership responsibilities, was $85,000–$120,000 (www.PayScale.com). According to the U.S. National Society of Professional Surveyors (www.nspsmo.org), entry-level salaries varied across the United States but averaged about $44,000 for a surveyor with a 4-year degree. Starting salaries could be higher or lower depending on the candidate's experience and other factors. Surveying technicians, who generally have 2-year degrees, had an average starting salary of about $25,000. Surveying technicians assist professional land surveyors by operating survey instruments and collecting information in the field and by performing computations and computer-aided drafting in offices.

3.1.3 QUALIFICATIONS

Although entry to geomatics/land surveying is possible from a range of disciplines, most employers prefer graduates who have completed a surveying (geomatics engineering) degree or who have at least taken an interest in this area by choosing surveying modules as part of their course. Examples of preferred degrees include geographic information science, surveying and mapping science, land/estate surveying, geography/physical geography, earth science, geology, environmental science, civil/structural engineering, mathematics, and physics. Entry with 2-year associate degree (or certificate) only is possible. Most associate degree holders would start as an assistant land surveyor or in a related role, such as digital mapping assistant/CAD technician. In some countries, it is also possible to enter surveying with training accessed through apprenticeship. They start as survey assistants and progress through a combination of work and study.

A master's degree in surveying or geomatics engineering can be helpful, but is not always essential. Postgraduate courses in more specialist areas are also available for those aiming to move into a particular area of the industry. These include subjects such as geodetic surveying, environmental management and earth observation, remote sensing, and geographical information science. Individual institutions provide details of courses and eligibility, but for entry to these types of courses, a first degree in a subject such as geomatics, engineering, geography, math, or physics is usually required. Potential candidates would need to show evidence of the following skills and qualities: knowledge of geographic information systems (GIS) and AutoCAD, and general IT skills; decision-making skills and the ability to work independently; oral and written communication skills; high levels of numeracy; map work orientation skills; accuracy, especially when using equipment; the capacity to identify problems quickly and to offer solutions; and the ability to conceptualize 2D and 3D information.

Internationally, there is lack of qualified geomatics engineers and many employers have found it difficult to recruit over the last few years.

3.2 ETHICS AND PROFESSIONAL CONDUCT

One of the phenomena that we all are concerned with, is the competition through bids for survey work. The danger is that submitting a low bid in order to get the contract may result in lowering the standard of the work provided to the client.

—Adler and Benhamu, 2007, 3

3.2.1 PRINCIPLES

In general, professional persons have many obligations to the public, not the least of which is the ethical obligation.

A profession is distinguished by certain characteristics, including: mastery of a particular intellectual skill, acquired by training and education; acceptance of duties to society as a whole in addition to duties to the client or employer; an outlook which is essentially objective; and rendering personal services to a high standard of conduct and performance.

(Allred, 1998, 154)

Surveyors fit very well in this definition—and their role poses certain ethical obligations on the exercise of professional duties. First, they have a duty to both the client and the public at large. Whether investigating a private boundary matter or performing a data collection survey, they require skills in order to determine the best information that a given situation allows. In performing their duty, it is important that surveyors be diligent, competent, impartial, and of unquestionable integrity in ensuring that the information that they provide is true, correct, and complete, to the best of their ability. They are often expected to give professional opinion based on facts that they have found for various outcomes or solutions to societal problems. They are the *fact finder and provider of geographic information.*

Second, sustainable development requires that their work is as much for the future as for the present; *their work has cumulative and long-term effects on future generations.* Information gathered is the basis for land management systems, which will enhance societal development; hence they have a duty to provide information on which knowledge is built. Many of the services are provided for society at large, and most information at some point in time becomes public information. That information may be used for purposes different from the purpose for which it was initially gathered.

The core principles upon which the FIG based a code of ethics for the international surveying community included (Allred, 1998):

- First, to determine what are the values that surveyors believe in. Ethics are the application of values.
- A code of ethics should be concise and deal only with true ethical values and not with self-serving principles such as restricting competition and fees.
- A code of ethics should apply to all members of the surveying profession, both public servants and private practitioners.
- A code of conduct must be dynamic to accommodate changing practices and the application of the code to those practices.

In addition to the core principles, the FIG suggested the following four questions as an objective standard, a yardstick against which to measure each statement or ethical principle:

What would happen if everyone acted this way?
What are the consequences of my actions for all people?
Would I want someone else to act this way for me?
Could my conduct stand up to fully informed, public scrutiny?

The following statement by Professor Allan Ryan also appeared in the *Edmonton Journal* of September 24, 1997:

> If you would be ashamed to have your activities disclosed to the community, that should be a warning. It might be questionable ethically and illegal.

3.2.2 Rule Ethics and Social Contract Ethics

A code of ethics stems from the principle that basic rules can be used to establish right or wrong of actions. It is also possible to distinguish rule-based ethics from judgment-based or social contract ethics. For example, an organization may develop a set of regulations (rule-based) dealing with how to account appropriately for client monies, and a set of core values (judgment-based) that it expects its members to apply in their work. Rules are relatively easy for practitioners and as one author puts it, "There is no room for judgement or opinion as to the correct solution; there is simply a correct answer and anyone, anywhere, who follows the appropriate procedures correctly, will arrive at this answer" (Dabson et al., 2007, 7).

Social contract ethics stems from voluntary acceptance of rules or core values by a group of individuals and not from the imposition of a code by, say, an employer or association. The danger is that such rules are subject to different interpretations by different individuals in different circumstances. The following nine core values of the British Royal Institution of Chartered Surveyors illustrate a typical example of social contract principles:

1. *Act with integrity:* Never put your own gain above the welfare of your client or others to whom you have a professional responsibility. Respect their confidentiality at all times and always consider the wider interests of society in your judgments.
2. *Always be honest:* Be trustworthy in all that you do—never deliberately mislead, whether by withholding or distorting information.
3. *Be open and transparent:* Share the full facts with your clients, making things as plain and intelligible as possible.
4. *Be accountable:* Take full responsibility for your actions, and don't blame others if things go wrong.
5. *Act within your limitations:* Be aware of the limits of your competence and don't be tempted to work beyond these. Never commit to more than you can deliver.
6. *Be objective at all times:* Give clear and appropriate advice. Never let sentiment or your own interests cloud your judgment.
7. *Always treat others with respect:* Never discriminate against others.
8. *Set a good example:* Remember both your public and private behavior could affect your own and other members' reputation.
9. *Have the courage to make a stand:* Be prepared to act if you suspect a risk to safety or malpractice of any sort.

3.2.3 FIG MODEL

The FIG has a code of ethics for the global surveying community. This represents quite a challenge because the FIG has a global membership representing surveyors from many different cultures and religions, with diverse moral and ethical values. However, the basis for commonality regarding ethical principles is that, despite different religions and cultures having their own specific rules, the following two seem to be common around the world.

The Golden Rule: "Do unto others as you would have others do unto you."
The Hippocratic Oath: "Above all, not knowingly to do harm."

The Hippocratic Oath is normally associated with the medical profession, but it has universal application.

The FIG recommends the following code of conduct as the minimum to be expected of all professional surveyors.

1. In general, surveyors
   - Exercise unbiased independent professional judgment;
   - Act competently and do not accept assignments that are outside the scope of their professional competence;
   - Advance their knowledge and skills by participating in relevant program of continuing professional development;
   - Ensure that they understand the fundamental principles involved when working in new areas of expertise, conducting thorough research, and consulting with other experts as appropriate; and
   - Do not accept assignments that are beyond their resources to complete in a reasonable time and in a professional manner.
2. As employers, surveyors
   - Assume responsibility for all work carried out by their professional and nonprofessional employees;
   - Assist their employees to achieve their optimum levels of technical or professional advancement;
   - Ensure that their employees have proper working conditions and equitable remuneration; and
   - Cultivate in their employees integrity and an understanding of the professional obligations of surveyors to society.
3. When dealing with clients, surveyors
   - Avoid any appearance of professional impropriety;
   - Disclose any potential conflicts of interest, affiliations, or prior involvement that could affect the quality of service to be provided;
   - Avoid associating with any persons or enterprises of doubtful character;
   - Do not receive remuneration for one project from multiple sources without the knowledge of the parties involved;

- Preserve the confidences and regard as privileged all information about their clients' affairs; and
- Maintain confidentiality during and after the completion of their service.

4. When providing professional services, surveyors
   - Seek remuneration commensurate with the technical complexity, level of responsibility and liability for the services rendered;
   - Make no fraudulent charges for services rendered;
   - Provide details on the determination of remuneration at the request of their clients; and
   - Do not sign certificates, reports, or plans unless these were prepared and completed under their personal supervision.

5. As members of a professional association, surveyors
   - Do not enter into arrangements that would enable unqualified persons to practice as if they were professionally qualified;
   - Report any unauthorized practice to the governing body of the profession;
   - Refuse to advance the application for professional status of any person known to be unqualified by education, experience, or character; and
   - Promote the surveying profession to clients and the public.

6. As business practitioners, surveyors
   - Do not make false or misleading statements in advertising or other marketing media;
   - Do not, either directly or indirectly, act to undermine the reputation or business prospects of other surveyors;
   - Do not supplant other surveyors under agreement with their clients; and
   - Do not establish branch offices that purport to be under the direction and management of a responsible professional surveyor unless this is actually the case.

7. As resource managers, surveyors
   - Approach environmental concerns with perception, diligence, and integrity;
   - Develop and maintain a reasonable level of understanding of environmental issues and the principles of sustainable development;
   - Bring any matter of concern relating to the physical environment and sustainable development to the attention of their clients or employers;
   - Employ the expertise of others when their knowledge and ability are inadequate for addressing specific environmental issues;
   - Include the costs of environmental protection and remediation among the essential factors used for project evaluation;
   - Ensure that environmental assessment, planning, and management are integrated into projects that are likely to impact on the environment; and
   - Encourage additional environmental protection when the benefits to society justify the costs.

3.3   INDIVIDUAL AND TEAM RESPONSIBILITIES

3.3.1   REGARDING YOUR WORK

Morse and Babcock (2007, 393):

> Doing an exceptional job on a minor assignment is the best way to be recognized and
> assigned more important, more challenging, more satisfying work.

Do not wait for others; get things done. Just because you have asked or agreed with a
colleague, co-worker, technician, or vendor to carry out a task or provide something
you need does not mean that it will happen in a timely fashion. Show interest and
follow up from time to time. In other words, beware of progress; prompt (and prompt
again if necessary) to check on progress. Find other ways to get the work done if
necessary. Be understanding, but persistent, and learn to know the difference.

Go the extra mile—and hour. Reputations are not made on a 40-hour week. Tough
assignments and goals with deadlines will always demand uninterrupted blocks of
time that never seem to be available during the regular work time. Success generally
goes to those who put forth the extra effort and meet the deadlines. However, it is the
individual's responsibility to balance this against other values—time for families,
recreation to stay whole and renewed, service to community, and other investments
of time that are important.

Be visible and professional. Look for chances to help out with tasks that are
important to the assignment at hand. This could be as simple as offering to research
some information, making that important phone call, sending that important e-mail,
editing data, or preparing a PowerPoint presentation. In all you do keep in mind the
professional responsibility, that is, in dealing with clients or outsiders, you represent
your profession and organization.

3.3.2   REGARDING YOUR BOSS

Keep your boss informed. Know what responsibilities your boss or team leader has
delegated to you, and never let them be caught by surprise. If something is going wrong
or an assignment will be late, let the boss know. If you are given a job to do, complete
it, or if your initial effort convinces you that it is not worth doing, convey your feelings
or findings to the boss—don't let him or her continue to think that you are working on
it when you are not. However, do not communicate too much or unwanted trivia.

Make your boss's job easy. Your primary responsibility is to help your boss, company, or team carry out their responsibilities, so give top priority to what the boss
wants or what the company or team's common goal is. Learn to be creative and try
not to just come to your boss or leader with a problem—whenever possible state the
problem, the alternative solutions you have considered, and your suggested action.

3.3.3   REGARDING ASSOCIATES AND COLLEAGUES

Be respectful of others. Value your associates and colleagues, learn their individual
abilities, and respect diversity in culture or opinion. Notice how your more effective

colleagues interact and how they get things accomplished. Do your part of the job and keep other members in the loop and updated.

Learn teamwork and professional culture. Know the overall team responsibilities and get to know the team members and their individual strengths (and how they might be of benefit to the various aspects of the task at hand)—this is as much an individual responsibility as it is a collective responsibility. Again, do your part and look for chances to help out with tasks.

3.3.4 COMMUNICATING YOUR IDEAS

Communicate effectively. The important element of communicating is the *understanding* of the message. Depending on the circumstances, understanding of the message may be enhanced by verbal or nonverbal forms of communication. Similarly, understanding may be obstructed by prejudices, or by desire not to hear or believe what is actually being communicated. It may also be obstructed by distractions (noise), inattention, or error in decoding of the message (e.g., problem with language interpretations). As a communicator, it is your responsibility to ensure that the message is definitively being relayed without ambiguity. Explain yourself (over and over again if needed) and look for signs that the message is actually understood. Remember that feedback offers the same potential for misunderstanding as the initial transmission, but face-to-face feedback is enhanced by nonverbal communication. Also beware of the potential for misinformation through *rumor* (or *grapevine*) communication.

Listen effectively. When communicating verbally, the art of effective listening is as important as effective communication. Listen positively and attentively, allowing the speaker to make his or her point. Consider the speaker's attitude and frame of mind: Is (s)he an optimist or pessimist? Reliable or unpredictable? When in doubt, rephrase the speaker's words. Consider the nonverbal language as well: *vocal* (tone, stress, length, and frequency of pauses), *facial* (expression and eye contact), and *body language* (posture, gestures, and movement).

3.3.5 STAYING TECHNICALLY COMPETENT

Know and master the latest. Because of rapid technological advancements, try to stay generally knowledgeable about a wide range of technologies and methods in your field or specialty. Even after graduation, stay ahead of the game and don't quit learning. Stay competent by keeping up to date with the professional literature such as technical journals and bulletins, and networking through professional meetings and conferences. Research the latest information, know the cutting-edge technology and the latest market trends in your field. If necessary, consider continuing education through training and career development opportunities, that is, consider the need for lifelong learning.

Learning continues until death and only then does it cease.

—*Hsun Tzu (298–238 B.C.)*

Stay professional. It is your responsibility to know what you are professionally qualified to do and to then do it competently. Learn to know when a project or task is beyond your scope. You might learn that some aspects of a project might need the help of other experts (e.g., through collaboration), or, if needed, accept to learn through someone who is more qualified. For tasks that are within your discipline and you are qualified, exercise professionalism. For example, being licensed or certified gives you authority to bid for and execute projects within your discipline. Also, membership of professional societies will contribute to your wider knowledge about the field (e.g., through access to latest information and research, current issues affecting your profession, the current practice, etc.).

3.3.6   Managing Your Time

Time is a very democratic resource: The prince and the pauper both have exactly the same amount to spend in a day.
Being busy is simple..., but being effective is difficult.

—Morse and Babcock, 2007, 516

Since modern humanity never seems to have enough time for everything, planning things and setting priorities are important. If necessary, evaluate your personal work habits and manage or limit unnecessary time wasters in your schedule. Some of the most common *time wasters* that can negatively impact on your use of time are briefly identified next. No one can completely avoid them, but the problems can be minimized through good time-management practices.

1. *Inadequate, inaccurate, or delayed information:* To avoid this problem if you are in a leadership role, use effective communication techniques to define clearly what information is needed, why, and when it is needed by. On the other hand, if you have been asked to provide information, ensure that you understand what is being asked for, and if in doubt don't just assume or guess—ask for clarification.
2. *Ineffective delegation:* Example of ineffective delegation may be the case where you assign or delegate responsibility without accountability for the results. To counter this, learn how to not only assign tasks, but to delegate authority and still exact accountability for results.
3. *Telephone interruptions:* Make effective use of answering machine and electronic mail (e-mail). If you have constructed a sensible work schedule, limit the number of times you check your messages once you are already working. If you must check messages, it should only be for very compelling reasons.
4. *Unclear communication:* Unclear communication leads to misinformation. Avoid ambiguity and unclear communication through the use of effective oral and written communication. Be sensitive to cultural language differences and be mindful of the audience (receptor of your message)—ensure that the right message is understood.

5. *Crises:* Crises are part of life, and be prepared to handle the unexpected. Leave some degree of freedom in your schedule for the unexpected, and consider crisis management options that might be available to you.
6. *Lack of self-discipline:* This leads to *indecision, postponement,* and ultimately *leaving tasks unfinished.* To overcome this problem, set a suitable work plan, a regular one if the work takes longer than just a few of days. Even if not feeling ready yet, try getting started with shorter regular work intervals and progressively increase the work time as you build momentum. In other words, *start early (before feeling ready); beginning early can be difficult until it becomes a habit.* In the same vein, beware of the following obstacles to beginning early:
   a. *Procrastination:* "I'm very busy; I have too many things to do. I wait until deadlines are near then I can do most things all at once. With my busy schedule I don't have the luxury to start this early."
   b. *Perfectionism:* "I either do something well or not at all."
   c. *Elitist:* "If you are really bright, you just do it all in one sitting."
   d. *Blockers:* "The idea of preliminaries wouldn't work for me. I do better if I wait until I have to do something, then I do it fast."

Set priorities and follow through. For every project set goals and priorities, schedule the most important goals, and follow through as scheduled. Categorize your goals as either A (highest priority), B (lower priority), or C (desirable, but postponable), and the time that you estimate each will require. Schedule your most important goals first, and if it looks like it will require large blocks of time, decide which parts can be tackled first to prepare for the rest later. *Avoid putting second things first by default.*

Consider your energy cycle and your work environment. You should be aware that your energy level is not a constant, but varies from hour to hour (or time to time). Therefore, work with constancy and moderation. It is better to arrange brief work sessions with breaks than to arrange long fatiguing sessions that go beyond the point of diminishing returns. Equally important, make your work environment comforting and comfortable or take steps to minimize unavoidable discomfort. Noise, poor lighting, uncomfortable seating, and inadequate space may inhibit your best work. If working outside, consider countermeasures to extreme weather or poor working conditions.

BIBLIOGRAPHY

Adler, R. and M. Benhamu. 2007. Professional ethics for licensed surveyors—the proposal for a social contract. 2007 FIG Working Week.

Allred, G. K. 1998. Ethics for the global surveying community. *The Australian Surveyor,* 43(3): 153–159.

Dabson, A., F. Plimmer, S. Kenny, and M. Waters. 2007. Ethics for surveyors: What are the problems? 2007 FIG Working Week.

Estopinal, S. V. 2009. *A Guide to Understanding Land Surveys.* 3rd ed. New York: Wiley.

Morse, L. C. and D. L. Babcock. 2007. *Managing Engineering and Technology.* 4th ed. Upper Saddle River, NJ: Prentice Hall.

U.S. Bureau of Labor Statistics. 2010. *U.S. Department of Labor, Occupational Outlook Handbook, 2010–11 Library Edition, Bulletin 2800*. Washington, DC: Superintendent of Documents, U.S. Government Printing Office.

EXERCISES

1. Find a code of ethics for your local organization and discuss it in comparison with the FIG code of ethics.
2. Two students were giving a presentation on a group project that had been assigned to them as part of a senior class assessment toward their major. The opening speech went something like this:

   Student A: Welcome to our group presentation on X. This project was done collectively between B and myself. I will let my colleague (student B) give part of the talk, then I will pick up from there.

   Student B: The objective of this project was to conduct a survey network design in... (the student's mind goes blank and at this point he turns to his colleague)... A, where was our project located?

   Student A: Northeast of City Y.

   Student B: Then continues: The initial project data were converted into the required format using a software which I don't know but my colleague will tell you about it later on in this presentation.
   a. Are there any ethical issues with the above presentation? Explain.
   b. Discuss the situation in terms of the individual and team responsibilities.
   c. Is the above communication effective? Explain.
3. You have just started your first full-time job with a small surveying consulting firm. While working on a seven-leg closed polygon traverse, you notice that one of the angles was not measured, but keep quiet because you don't want to cause any trouble or look stupid. This is your third day on the job and you are starting to feel uneasy about the party chief who curses frequently, seems mildly aggressive, and appears to have a low opinion of educated surveyors. While on your way back to the office (the party chief is driving), the party chief tells you to sum up the measured interior angles in the traverse. You do this and say 843°18′22″. (Now you should remember what the sum should be.) The party chief next tells you to write an angle 56°41′24″ into the field book where the missing angle should have been written. You are then told to adjust the angles, work out the azimuths, and start a preliminary traverse computation. Which of the following is your most ethically correct response?
   a. Do as you are told because you really need this job.
   b. Tell the party chief, "I quit, drop me off immediately, I'll find my own way home."
   c. Wait until you get back to the office and then talk to the party chief's supervisor alone, explain the situation and ask for guidance.
   d. Immediately ask the party chief why the data have to be made up.

4 Policy, Social, and Environmental Issues

4.1 POLICY ISSUES

Most laws are statements of policy broadly outlining what the law prohibits or allows, as well as the penalties for noncompliance. Understanding the current policy issues that affect your profession is of utmost importance. Land surveys should be designed and performed in such a way that they do not contravene governmental regulations and broader policies.

4.1.1 PROFESSIONAL QUALIFICATIONS

Every land surveyor should be familiar with the government act or regulation defining the profession and pertinent matters such as registration and licensure. The common law is that a land survey may only be performed and signed by a licensed (or registered) Professional Land Surveyor. The qualification requirements to become a Professional Land Surveyor are discussed in Chapter 3. Licensure requirements vary from state to state (and country to country). For example, in the U.S. state of New Jersey, an individual must have a 4-year college degree in surveying, 3 years or more of practical experience, and pass a 16-hour written examination administered by the New Jersey Division of Consumer Affairs in order to meet the qualifications for licensure in that state. Once licensed, the Professional Land Surveyor must obtain 24 hours of continuing education credits every 2 years to maintain active status. In every country, the government or other entity (such as a Board of Registration) establishes a strict code of regulations for licensing and outlines the procedures and requirements to become licensed. On behalf of the government, the board implements a public policy to license land surveyors.

4.1.2 ACCESS TO PUBLIC INFORMATION AND RECORDS

In the United States, it is a policy of the government that land survey records created by licensed surveyors for purposes such as mapping, building and property line, are accessible to the general public. The policy is implemented through the Office of Public Land Surveys, which has the duty to collect, preserve, and index such public land survey records. This serves to reduce research time, the likelihood of conflicts between surveyors about corner positions, and the chance of costly lawsuits over boundary disputes. For a land survey to be commissioned, a licensed surveyor must survey the property and publish the record for public and private viewing.

DOI: 10.1201/9781003297147-6

The land surveys must meet the requirements specified by the American Land Title Association (ALTA).

The land survey records are available for public access and can be viewed online using key information about the property such as the township, concession, and lot or block number. You can also use the address if this is the only information you have, but you will still need to use the address to find the township, concession, and lot or block number.

It is also important to note that the public records policy that pertains to land surveys will vary from state to state (and country to country). Here is an example of a public records policy statement by the State of Wisconsin Board of Commissioners of Public Lands:

PUBLIC RECORDS POLICY

You have a right to inspect and copy certain records under Wisconsin Public Records Law. This notice is posted pursuant to §§ 19.34 of the Wisconsin Statutes.

PROCEDURES TO FOLLOW TO REQUEST ACCESS TO RECORDS

1. A request for access to a public record may be made orally or in writing and should be directed to the Records Custodian or Deputy Custodian believed to have the records requested. The request for access to a public record must reasonably describe the record requested and must be reasonably limited as to the subject matter and/or length of time represented by the record.
2. Requests for access to, and inspection of, any public records may be made at the Office of the Executive Secretary during the hours of 8:00 am to 4:30 pm M-F. Requests for access to, and inspection of, District Office land records may be made at the District Office during the hours of 8:00 am to 10:00 am on Tuesdays. In addition, appointments may be made upon 24 hours' notice. For best service at either office, appointments are highly recommended but not required.
3. A request may be denied if the particular document does not exist, is excepted by state law from the definition of a public record, is exempted from public access by state or federal law, or when the public interest against disclosure outweighs the public interest in disclosure.
4. Reasonable restrictions may be imposed on the manner of access to an original record if the record is irreplaceable or easily damaged.
5. The office may impose a fee for searching and copying records pursuant to Wisconsin Statute 19.35(3). The copy fee is $0.25 per page for a standard photocopy but the office may charge more for copies of oversize documents in accordance with Statute 19.35(3). Search fees of $35 per hour may be imposed if the cost of the search is $50 or more. The office may require a prepayment of any fees if the total copying or search fees exceed $5.

4.1.3 MANDATORY FILING REQUIREMENTS AND FEES

All public land surveys must be filed with an Office of Land Surveys in the state in which they were done. Each state has specific laws and it is important to be familiar with those laws especially the filing fees and miscellaneous, including the consequences for noncompliance. There may be different requirements, for instance, fees for a subdivision application, fees for filing an original report of survey, fees for filing of reports relating to altered or destroyed markers, and fees for reproduction of survey records. The following excerpts are taken from the Kansas Statute (KSA 58-2011) to serve as an example:

> ... (a) Whenever a survey originates from a United States public land survey corner or any related accessory, the land surveyor shall file a copy of the report of the completed survey and references to the corner or accessory with the secretary of the state historical society and with the county surveyor for the county or counties in which the survey corner exists. If there is no county surveyor of such county, such report shall be filed with the county engineer. If there is no county engineer, such report shall be filed in the office of the county road department. Reports filed with the secretary of the state historical society may be filed and retrieved using electronic technologies if authorized by the secretary. Such report shall be filed within 30 days of the date the references are made. At the time of filing such report with the secretary of the state historical society, the land surveyor shall pay a filing fee in an amount fixed by rules and regulations of the secretary of the state historical society. Fees charged for filing and retrieval of such reports may be billed and paid periodically.
>
> ... (b) Any person engaged in an activity in which a United States public land survey corner or any related accessory is likely to be altered, removed, damaged or destroyed shall have a person qualified to practice land surveying establish such reference points as necessary for the restoration, reestablishment or replacement of the corner or accessory. The land surveyor shall file a reference report with the secretary of the state historical society and with the county surveyor for the county or counties in which the survey corner exists. Such report shall be filed within 30 days of the date the references are made. At the time of filing such report with the secretary of the state historical society, the land surveyor shall pay a filing fee in an amount fixed by rules and regulations of the secretary of the state historical society.
>
> ... (d) Failure to comply with the filing requirements of this section shall be grounds for the suspension or revocation of the land surveyor's license.
>
> ... (g) The failure of any person to have a land surveyor establish reference points as required by subsection (b) shall be a class C misdemeanor.

4.1.4 BEST PRACTICE GUIDELINES, RULES, AND PROCEDURES

Rules and guidelines relating to matters of standards, best practices, methods, procedures, and appropriate use of technology are the most popular means to advance policies on the conduct of projects and use of geographic information. Mostly, manuals and guidelines are designed to guide the members of a profession or employees of a particular organization in the conduct of projects. However, that information is also usually made available for public use as well. Various governing bodies and committees are usually put in charge of drafting such manuals and guidelines on the basis of wider consultation among experts and/or members of the profession. Guidelines exist for things such as establishment and use of national reference systems for project control, procedures for achieving various standards, procedures for procurement of professional services, and general best practice policies for the profession. A few examples are as follows (not in any particular order, and many others can be included internationally):

1. Manual of Surveying Instructions by the U.S. Bureau of Land Management
2. NGS Policy on Submitting Data for Inclusion into the NSRS
3. NGS User Guidelines for GNSS Real-Time Positioning
4. NGS Network Guidelines for New and Existing Continuously Operating Reference Stations
5. BLM Guidelines for Cadastral Surveys Using Global Positioning System Methods
6. BLM Guidelines for the Use of Global Navigation Satellite Systems (GNSS) in Cadastral Surveys
7. BLM Standards for Positional Accuracy for Cadastral Surveys Conducted Using Global Navigation Satellite Systems (GNSS)
8. ASPRS Accuracy Standards for Digital Geospatial Data
9. ASPRS Guidelines for Reporting Vertical Accuracy of LIDAR Data
10. ASPRS Guidelines for Reporting Horizontal Accuracy of LIDAR Data
11. ASPRS Standards for Large-Scale Maps
12. CALTRANS Surveys Manual
13. FGCS Standards and Specifications
14. USGS GIS Guidelines
15. ASPRS Guidelines for the Procurement of Professional Services
16. ASPRS Guidelines for Procurement of Professional Aerial Imagery, Photogrammetry, LIDAR and Related Remote Sensor-Based Geospatial Mapping Services
17. ALTA/American Congress on Surveying and Mapping (ACSM) publications (e.g., Minimum Standard Detail Requirements and Accuracy Standards for ALTA/ACSM Land Title Surveys)
18. NOAA Guidelines for Establishing GPS-Derived Ellipsoid Heights
19. FGCS Specifications and Procedures to Incorporate Electronic Digital/Bar-Code Leveling Systems
20. FGCC Standards and Specifications for Geodetic Control Networks
21. FIG Guidelines on the Development of a Vertical Reference Surface for Hydrography
22. USACE Engineering Regulations and Manuals
23. Guidelines for GPS Surveying in Australia
24. Guidelines by international organizations such as IGS and IAG
25. ASCE Standard Guidelines for the Collection and Depiction of Existing Subsurface Utility Data.

Let us expand on some of these, in turn:

1. *BLM Guidelines for Cadastral Surveys Using Global Positioning System Methods:* The USDA Forest Service and the Bureau of Land Management developed Standards and Guidelines for Geodetic Control Ties and Cadastral Surveys Using GPS and Other Methods to set a policy for compliance with special instructions for official government surveys. The purpose

is to provide direction for reporting geodetic control ties to the Public Land Survey System (PLSS) and making Cadastral Survey Measurements when using Global Positioning System (GPS). The policy includes the use of the NGS Online Positioning User Service (OPUS) as an acceptable method for georeferencing official surveys. Here is an excerpt of the guidelines:

Our goal is to provide the highest quality of measurements possible for geodetic ties from the National Spatial Reference System (NSRS) to a minimum of two monuments of an official government survey. The field surveyor will determine the applicable data collection method to be used during the course of the field survey. The purpose of the control ties is to increase the accuracy of the Geographic Coordinate Data Base and to densify the BLM GPS control network for future government surveys.

-Types of control stations-

The following is a list of types of control stations that can be used for georeferencing official surveys. They are listed from the highest to the lowest order of expected accuracy. To ensure the highest quality data for project control use the best available control near your project.

1. The NGS maintains Continuously Operating Reference Stations (CORS) that can be used to differentially correct static GPS data. By NGS definition, the local and network accuracy values at CORS sites are considered to be infinitesimal, that is, to approach zero or have no error. The NGS has developed an Internet-based method of differentially correcting static GPS automatically for you. This is done using their OPUS site, to correct static GPS against multiple CORS stations.

2. The highest order accuracy of set monuments of the NSRS stations is High Accuracy Reference Network (HARN) stations. The NGS refers to their control station monuments as marks. The HARN marks are established using GPS measurement methods. These are A and B order HARN stations; which are 1:10,000,000 and 1:1,000,000 precision stations, respectively. The control station coordinates are published by the NGS.

3. First, second-, and third-order stations generally established using conventional instruments (optical and laser instruments) are 1:100,000, 1:50:000, and 1:10,000 precision stations, respectively. There are some circumstances where NGS marks have been observed with GPS and blue booked, and are not part of the HARN. These lower order GPS stations generally fit well when constrained in GPS networks. In this memo conventional survey and survey methods will refer to surveys that use optical and laser instruments, for triangulation, trilateration, and traversing. The control station coordinates for non-HARN NSRS marks are published by the NGS.

4. Other control stations include BLM PLSS or GPS control monuments with geographic coordinates or local control stations established by private land surveyors or other government agencies. The origin and

quality of the geographic positions of both BLM and local control stations must be verified before use. The positional quality should be consistent specifications of survey-grade GPS equipment, meaning the centimeter accuracy range. These types of station coordinates are not generally published.

If using NSRS marks with geographic coordinates derived by conventional survey methods, the surveyor may have to tie to more than two NSRS marks to confirm the relative accuracy of published control coordinates in an adjusted network.

-Methods of data collection-
The following is a list of methods for data collection that can be used for georeferencing official surveys. They are listed from the highest to lowest in order of expected accuracy of network adjustment results.

1. Use survey-grade dual-frequency GPS receivers. Make static observations tying survey monuments to NSRS HARN or CORS stations of the best available accuracy. If using a CORS within a reasonable distance range of your project download the CORS file needed for your specific observation times from the Internet, to correct files that contain less than 2 hours of data. The accuracy of control coordinates should be confirmed using least squares analysis of derived GPS baselines in an adjusted network. The network should include ties to a minimum of two NSRS marks or CORS.

2. Use survey-grade dual-frequency GPS receivers. Make static observations tying survey monuments to NSRS CORS stations. A minimum of 2 hours of data should be collected, and 4 hours is preferred by NGS to use OPUS. Submit collected data to OPUS as per instructions in the attached PDF file named OPUS directions.pdf. Particular attention should be paid to the reported antenna heights as incorrect HI measurements give wrong answers. The accuracy of control coordinates should be confirmed by comparing GPS coordinate returns from NGS OPUS program. A minimum of two independent OPUS sessions are needed to derive and check geo-coordinates for each survey monument.

3. Use survey-grade GPS receivers. Make real-time kinematic measurements directly from a NSRS station, or a BLM or local survey control station established using static GPS methods from NSRS marks or CORS. Use the best available accuracy control stations near the project. Control coordinates should be verified using least squares analysis of GPS baselines in an adjusted network. The network should include ties to a minimum of two control stations.

   ... When working in a remote area, using OPUS may be the most economical method for acquiring geodetic coordinates.

4. Use conventional survey equipment. Use conventional survey methods to tie to a NSRS, BLM, or local control station. Use the best available accuracy control stations near the project. Control coordinates should

be confirmed using least squares analysis of survey measurements in an adjusted network. The tie should be a closed traverse to the control station and back. A copy of IM 83-55, Guidelines for Cadastral Control, is available from the Idaho State Office Geodesist summarizing how to properly make conventional geodetic survey ties.

5. Use precision lightweight GPS receivers or other code or carrier-phase resource-grade GPS receivers that can be differentially corrected. This method can be used for projects that encompass one PLSS section or less in area. This method is a last resort method for obtaining geographic coordinates of survey monuments. Consult with your field section chief prior to using this method.

2. *CALTRANS Surveys Manual:* The surveys manual by the California Department of Transportation (CALTRANS) is yet another example of guidelines outlining the policy for surveying projects. Again, the manual is not binding to every surveyor except those who opt to use it or are otherwise required by the client or employer to follow its guidelines. It is binding to a CALTRANS employee doing a project for or on behalf of the CALTRANS. At the time of writing this book, the manual contained a total of 14 chapters but only 12 were publicly accessible. Here are excerpts of some of the chapters that fit the context of this book:

Chapter 2: Safety. This section of the manual is intended to: (1) provide safe operating procedures, guidelines, and practices, specific to CALTRANS surveying operations; and (2) supplement the policies, procedures, and practices set forth in the CALTRANS Safety Manual. The CALTRANS Safety Manual provides detailed instructions for managers, supervisors, and employees to assist them in their individual efforts to conduct CALTRANS business in a safe and healthy manner consistent with current law, rule, and technology.

Chapter 3: Survey Equipment. This chapter provides policy, procedure and general information on procurement, control and maintenance of survey equipment, tools, and supplies.

Chapter 4: Survey Datums. Universally accepted and used common survey datums are essential for the efficient sharing of both engineering and Geographic Information System (GIS) data with CALTRANS partners in developing and operating a multimodal transportation system.

All engineering work (mapping, planning, design, right-of-way engineering and construction) for each specific CALTRANS-involved transportation improvement project shall be based on a common horizontal datum.

The horizontal datum for all mapping, planning, design, right-of-way engineering, and construction on CALTRANS-involved transportation improvement projects, including special funded State highway projects, shall be the North American Datum of 1983 (NAD83), as defined by the National Geodetic Survey (NGS). The physical (on-the-ground survey station) reference network for the NAD83 datum for all CALTRANS-involved transportation improvement projects shall be the California High Precision Geodetic Network (CA-HPGN) and its densification stations (CA-HPGN-D).

Chapter 5: Classifications of Accuracy and Standards. All surveys performed by CALTRANS or others on all CALTRANS-involved transportation improvement projects shall be classified according to the standards shown on the chart in Figure. Standards shown are minimum standards for each order of survey.

In addition to conforming to the applicable standards, surveys must be performed using field procedures that meet the specifications for the specified order of survey. Specifications for field procedures are provided in Chapter 6, "Global Positioning System (GPS) Survey Specifications"; Chapter 7, "Total Station Survey System (TSSS) Survey Specifications"; and Chapter 8, "Differential Leveling Survey Specifications." Survey accuracy standards are meaningless without corresponding survey procedure specifications. Without the use of proper specifications, chance and compensating errors can produce results that indicate a level of accuracy that has not been met.

The order of accuracy to be used for a specific type of survey is listed on the chart in Figure. Tolerance requirements for setting construction stakes are provided in Chapter 12, "Construction Surveys." Tolerance requirements for collecting data are provided in Chapter 11, "Engineering Surveys."

Chapter 13: Photogrammetry. This chapter is to be used for Department-involved transportation improvement projects, including special funded projects. It shall be used by all Department employees, local agencies, and consultants performing photogram-metric tasks. It describes a statewide model of responsibilities and workflow. Unique circumstances in a project may warrant deviations from this model.

The following are factors to consider when deciding to use photogrammetry:

- Photogrammetry is a cost-efficient surveying method for mapping large areas.
- Photogrammetry may be safer than other surveying methods. It is safer to take photographs of a dangerous area than to place surveyors in harm's way.
- Photogrammetry provides the ability to map areas inaccessible to field crews.
- Photogrammetry creates a photographic record of the project site (snapshot in time).
- Photogrammetry produces useful digital products such as orthophotos.
- Photogrammetry produces electronic terrain models.

    Photogrammetry may not be appropriate under the following conditions:

- The accuracy required for a mapping project is greater than the accuracy achievable with photogrammetric methods.
- The scope of the work is not large enough to justify the costs of surveying the photo control and performing the subsequent photogrammetric processes. However, when unsafe field conditions are encountered, safety shall hold a higher weight than cost in the decision process.

3. ASPRS Guidelines for Procurement of Professional Aerial Imagery, Photogrammetry, LIDAR and Related Remote Sensor-Based Geospatial Mapping Services. Available online at www.asprs.org/guidelines, the intent of these guidelines is to provide public agencies, researchers, private entities, and other organizations with a resource they can use as a guide to help determine the best approach and methodology for procuring photogrammetry and related remote-sensor-based geospatial mapping services. However, the American Society for Photogrammetry and Remote Sensing (ASPRS) has cautioned that this guidelines document is not intended to establish any laws, rules, or regulations. Rather, it has set forth these guidelines as recommendations "for procurement methods which are, in the opinion of ASPRS, the best and most appropriate means of procuring professional photogrammetry, remote sensing services and related geospatial mapping services."

The guidelines were prepared by a committee, which included representations from the ASPRS Professional Practices Division, ASPRS members from state and federal government, the Management Association for Private Photogrammetric Surveyors, and the ACSM. Prior to approval by the ASPRS Board of Directors in August 2009, the development process spanned more than 3 years and included numerous public presentations, comment periods, and draft publications.

Here are some elements of the guidelines:

"... Procurement methods should consider potential impacts to the intended end application."

"Qualifications-based procurement methods are the most appropriate means to ensure that public interests are best represented when procuring professional photogrammetry, remote sensing and related sensor-based geospatial mapping services..."

ASPRS considers professional geospatial mapping services to be those geospatial mapping services that

1. Require specialized knowledge derived from academic education, on-the-job training, and practical experience;
2. Produce mapping deliverables and information where there is an expectation or representation of geospatial or thematic accuracy;
3. Require independent judgment, ethical conduct, and professional expertise to ensure that the resulting products derived from these services represent the best interests of the client and public; and
4. Could potentially affect public welfare or result in harm to the public if not performed to professional standards.

   ASPRS recommends the following guidelines be applied to any procurement method that does not adhere to the preferred process outlined by the Brooks Act.
1. Qualifications should always be the primary selection factor.
2. Qualifications rankings should not be influenced by cost.
3. The scope of work must be well defined and have been developed by a professional who has extensive knowledge of the work to be performed

and is qualified to ensure that the scope of work will best serve the client's interests.

4. Projects that have a significant element of design, and where the service providers' professional judgment is relied on to develop the scope of work, methodology, or approach, should always use Brooks Act QBS and should not include cost as selection criteria.

5. A registered, certified, or otherwise qualified professional with specific knowledge or expertise with the services being procured (either on the client's staff or hired as a consultant) should have a significant role in the review of both the technical proposal and any cost proposals in order to ensure that the work best meets the end-user and public interests.

6. If cost data are to be considered in the selection process, it should be submitted separately and considered only after firms are ranked based on qualifications.

4.1.5 LAND (DEVELOPMENT) POLICIES

Land policy is a conscious action to bring about an optimal use of land as well as of a socially just distribution of landownership and of income from land. National land policies are as diverse as the many different countries that exist worldwide. They address matters of land title policies, that is, laws relating to land ownership and registration and land dispute settlement mechanisms. It is important to understand that different states and nations have different policies to define legal land ownership, and rules that regulate those matters which, arising from the interests of society, require regulation in relation, to land transactions, parcels of land, property relations, and kinds of use. Some aspect of this is discussed further in Section 4.2.

In order to ensure proper land management, it is necessary to maintain a good land information system. It is important to know how much land is occupied by whom and for what purposes, and how much land is still left for further development. A policy of keeping good land records is essential for planning purposes and for protecting existing land rights. It creates a condition conducive to sound project planning and design.

Knowing the existing laws for a given land policy will help in terms of planning for things like land records research and how to go about planning for and conducting a land survey. While in countries like the United States this is not a problem and it may be a somewhat straightforward business, some countries have poor land records for various reasons. Some of those reasons include

1. Most land is occupied under traditional (customary) law and is not recorded or registered

2. Unregistered statutory allocations (e.g., national parks, forest reserves, etc.) result in many such areas being encroached upon and alienated

3. Unregistered government allocations (e.g., land under government buildings) also encourage encroachment

4. Slow process in the preparation of documents evidencing rights of land ownership resulting in some areas remaining unregistered for many years and, because of this, a number of double allocations occur, resulting in many land disputes.

The last point is especially important to underscore because only surveys made by licensed and registered surveyors are legal and acceptable in courts of law. Having registered interests in land, proper record-keeping systems, and properly laid out land dispute resolution mechanisms will enhance project planning processes. Nevertheless, understanding the existing situation and culture (such as in countries with poor land records) is equally important.

4.1.5.1 Local Land Use Regulations

4.1.5.1.1 Example 1: Land Use Zoning Regulations

In the United States, *land use zoning* is a local or county law (ordinance) setting up everything in the town or county into what are called districts. Some states allow agricultural land to be exempt. A district has two sets of regulations connected with it:

1. *Permitted uses:* This simply means permitted land uses and building uses in a given land development district. In this case, a list of permitted uses might include single-family residences, day care centers, group homes, offices, landfills, factories, places of worship, and so forth. Each of these might have many qualifiers attached to it. For example, the term might be "offices of less than 5,000 square feet," or "single-family detached homes," or "factories meeting the performance standards set forth in Section 4.1 of this ordinance."
2. *Regulations:* This commonly includes the minimum square footage of the lot on which the land use sits, any required feet of front yard setbacks from the street, minimum rear and side yard setbacks of the building from the property line, number of parking spaces required, percentage of the lot that may be occupied by the building, and whether any signs are permitted. The variations are endless.

4.1.5.1.2 Example 2: Subdivision Regulation

Another type of law that often gets confused with zoning is the subdivision regulation. When you have a chunk of land and you want to make it into two or more parcels of land so you can sell some of it, that is called subdividing. You might be dividing your 400-acre farm into four 100-acre farms, and if your county had a subdivision regulation, you might have to apply to subdivide. So don't let the common meaning of subdivision make you think that subdivision regulations only pertain when you want to build tract houses.

The subdivision regulation goal is to make sure that each lot (or land parcel) is reasonably uniform, that no awkwardly shaped lots are formed, and that each lot can be served by utilities and roads. It also establishes and publishes the surveying markers that will be used by professional surveyors to measure land and establish where each lot to be sold lies.

4.1.5.1.3 Example 3: CC&Rs, Covenants, and Deed Restrictions

CC&Rs stands for covenants, conditions, and restrictions, another type of land use regulations. For instance, a covenant might say "this property must be forever green space." "This property can never be used as a cemetery." "This property must always be used as a farmer's market." "There can never be any public service of alcohol on this property."

4.1.6 ENVIRONMENTAL POLICY AND REGULATIONS

A project might require that you comply with environmental planning and natural resource protection laws. If that is the case, find the appropriate land use laws and regulations or consult with an expert in this area. Examples of such regulations include those pertaining to governmental power and authority to zone land. Other examples might include conservation laws prohibiting vegetation clearance (which might impact on access to a project site or the method of choice for conducting the land survey) and laws stipulating preferred class of survey. Let us consider two specific examples:

4.1.6.1 Example 1: Environmental Land Use Restrictions

Given next are two excerpts from the "Regulations for Environmental Land Use Restrictions" in the U.S. state of Connecticut (January 30, 1996). The emphasis has been added to highlight the aspects of the regulations that might have an impact on project planning. The first excerpt indicates where to search for (and find) evidence of such land use regulations. The second excerpt identifies the required class of survey for such lands—especially important to know if you are the one to conduct the survey:

> ... Except as otherwise provided by section 22a-133o of the General Statutes, no environmental land use restriction shall be effective unless and until it has (1) been submitted to the Commissioner for his review and approved by him as evidenced by his signature on the original of the instrument setting forth such restriction; and (2) been *recorded on the land records in the municipality in which the subject parcel is located...*
>
> When submitting a proposed environmental land use restriction to the Commissioner for his review and approval, the owner of the affected parcel of land shall simultaneously submit:
>
> ... (2) *a Class A-2 survey of the parcel or portion thereof* which is the subject of the proposed environmental land use restriction;...

4.1.6.2 Example 2: Specially Protected Areas

There are also legislations governing "Specially Protected Areas such as Forests and National Parks." Here we will consider one international example in which the government has authority to vary boundaries in order to protect such areas, namely, the "Forest Act, Chapter 385, and Wildlife (Conservation and Management) Act, Chapter 376, Laws of Kenya" (Aketch, 2006):

> The Forest Act empowers the relevant minister to declare unalienated government lands to be forest areas, and *to vary the boundaries* of such forest areas. Further, the minister may declare a forest area or some part of it to be a "nature reserve," for purposes of preserving the flora and fauna found therein. The act establishes the office of the Director of Forestry, and gives it responsibilities such as issuing exploitation licenses. Conversely, the Wildlife Act provides for the protection, conservation and management of wildlife. It gives the responsible minister powers similar to those granted by the Forest Act. Thus the minister may declare a given area to be a national park, game reserve, or sanctuary. Further, the minister *may similarly vary the boundaries* of such special areas. The act is implemented by the Kenya Wildlife Service, which is a state corporation.

4.2   SOCIAL AND GLOBAL ISSUES

There are certain projects that will require serious consideration of social global issues. For instance, consider the following examples:

1. A project establishing geodetic GPS/Global Navigation Satellite Systems (GNSS) CORS networks for the U.S. forces in Iraq and Afghanistan
2. A mapping project in Saudi Arabia by a U.S.-based company
3. A mapping project in Africa for the United Nations or World Bank
4. A GIS mapping project for the United Nations Environment Programme in South East Asia
5. A road construction survey project in Libya for a private U.S.-based multinational company
6. A GPS/GNSS CORS infrastructure development project in an aid-recipient country as part of a bilateral aid agreement.

There are myriad issues to be considered when dealing with such projects. For example, in a global context, you might want to consider who the project stakeholders are, cultural and language issues, bureaucracies, ethical considerations, contracting laws and regulations, risk issues and risk management, economic infrastructure (e.g., transportation and utilities), worldwide land registration systems, environmental issues, and so forth.

Generally, the *timing* or *duration* (and hence *cost*) of a project can be affected by the following issues if it is located in countries that are described as "third world" or "developing" countries:

1. Government planning of the economy, and frequently, government operation of utilities
2. Poor or lack of economic infrastructure such as transportation and utilities, land registration and record-keeping systems
3. A shortage of skilled workers, professionals, and support services

4. Cultural differences influencing attitudes toward work
5. Climatic conditions such as harsh summer temperatures limiting human output
6. Remoteness of a project location and lack of communication facilities
7. Communication timing problems, for example, due to 9-hour time difference and 4-day weekend in some countries (Thursday and Friday in Saudi Arabia, Saturday and Sunday in the West)
8. Government bureaucracy leading to delays in obtaining permits, documents, and critical data for projects
9. Religious law and culture: reduction in efficiency during the fasting month of Ramadan due to Islamic law, time and facilities for prayer five times daily, and restrictions on women's work and prohibition of their driving
10. Expatriate problems in hiring (lead time, restrictions on nationality, religion, and sex; differential pay scales, languages, cultures, values, motivations, and work methods).

Despite this list, each part of the so called third world will have its own set of problems. To be successful in an unfamiliar environment, be very careful to become familiar with the local culture, politics, and people and learn to operate under these constraints.

Let us further examine three of the issues mentioned earlier.

4.2.1 PROJECT STAKEHOLDERS

Stakeholders are the primary and secondary users of new information to be collected and analyzed as a result of a project. It is necessary to consult with potential stakeholders (and donors, financiers) in the design of the project. If, for instance, the project is as a result of U.S. government aid to the government of Afghanistan, here are examples of some potential questions to consider:

- Are the desired outcomes of the project part of a national policy by the aid-recipient government?
- Who (or which agency) is mandated by the aid-recipient government to collect and create mapping data on a national level?
- Are there any principal risks that may affect the achievement of outputs and outcomes of the project?
- Is the project part of a long-term future goal, for instance, to enhance internal technical capacity of the recipient government?
- Is it the policy of the recipient government that information and mapping products generated from the project will be widely distributed? Who will be the primary beneficiaries of the information?

Consider all the stakeholders and their needs, that is, the value, applications, and uses they will attach to the outcomes of the project.

4.2.2 BUREAUCRACY AND ETHICAL CONSIDERATIONS

A project with a global context will certainly need considerations of ethical challenges. Morse and Babcock (2007, 433) summarized these considerations when dealing with project management in developing countries:

> Because cultures vary so much, engineering and managerial work in developing countries may involve ethical decisions that would not present themselves in the United States. For example, bribery of public officials may be a way of life and may be necessary to get permits issued or spare parts released from customs. Minor officials may be paid so little that to maintain a decent livelihood, they count on supplements that would clearly be illegal in the United States. Should the American plant manager provide such "grease" (or baksheesh or whatever the local term is)?...
>
> ... Should a U.S.-owned plant just maintain levels of pollution control and plant safety consistent with local requirements, or should higher levels required in U.S. plants be maintained, at higher cost?

There are no simple solutions to such ethical problems. And the challenges will be varied according to the location and nature of the project. For example, for a mapping or land survey project, the problem might be difficulty in access of local land records due to government control and bureaucracy.

4.2.3 WORLDWIDE LAND REGISTRATION SYSTEMS

Another aspect that might be important from a global perspective is the worldwide diversity of cadastral (land registration) systems. One of the best sources to understand the land registration systems is Professor Stig Enemark, a recognized international expert in the areas of land administration systems and president of the International Federation of Surveyors, FIG 2007–2010. He gave the following definitions in a 2009 conference paper:

> Cadastral Systems are organized in different ways throughout the world, especially with regard to the Land Registration component. Basically, two types of systems can be identified: the Deeds System and the Title System. The differences between the two concepts relate to the cultural development and judicial setting of the country. The key difference is found in whether only the transaction is recorded (the Deeds System) or the title itself is recorded and secured (the Title System). The Deeds System is basically a register of owners focusing on "who owns what" while the Title System is a register of properties presenting "what is owned by whom." The cultural and judicial aspects relate to whether a country is based on Roman law (Deeds Systems) or Germanic or common-Anglo law (Title Systems). This of course also relates to the history of colonization.
>
> The sources of different land registration systems worldwide are shown indicatively in Figure 4.1. Minor regional discrepancies may occur, but the map does indicate the overall distribution of three types of land registration systems throughout the world.

Land registration systems around the World

French system
German system
English system
Torrens system

Diagonal indicates a mixed system

Deeds system (French): A register of owners; the transaction is recorded – not the title.
Title system (German, English, Torrens): A register of properties; the title is recorded and guarantied.

FIGURE 4.1   World map of land registration systems (Enemark, 2009).

4.3   ENVIRONMENTAL ISSUES

The FIG Code of Professional Conduct recommends that surveyors as resource managers:

- Approach environmental concerns with perception, diligence, and integrity;
- Develop and maintain a reasonable level of understanding of environmental issues and the principles of sustainable development;
- Bring any matter of concern relating to the physical environment and sustainable development to the attention of their clients or employers;
- Employ the expertise of others when their knowledge and ability are inadequate for addressing specific environmental issues;
- Include the costs of environmental protection and remediation among the essential factors used for project evaluation;
- Ensure that environmental assessment, planning, and management are integrated into projects that are likely to impact the environment; and
- Encourage additional environmental protection when the benefits to society justify the costs.

4.3.1   SUSTAINABLE DEVELOPMENT

The principle of *sustainable development* permits opportunities for economic growth but at the same time demands protection of the environment for the benefit of future generations. Human settlements and urbanization bring about many types of environmental change—activities such as deforestation, mining, and intensive use of rural land cause environmental degradation. Be prepared to explain how environmental sustainability is relevant to the client as well as to society at large. For example,

1. Pollution and environmental degradation affect human health and mental health.
2. Clean energy and sustainable building practices could bring new industries for economic development.
3. Reducing your carbon footprint as an individual, neighborhood, community, or government could help slow global climate change.
4. Energy demand, together with supply, determines energy cost. In turn, energy costs impact where people can live and work—important community development considerations.

Think of the following as "debatable" examples of how surveyors can contribute to sustainable development (add to the list if you can!):

- Whenever presented with options and if practical and cost-effective, use satellite-based technologies for mapping instead of aerial mapping or ground-based methods that rely on the use of fossil fuel.
- Whenever practical and cost-effective, support the use of existing continuously operating GPS/GNSS networks infrastructure for project control instead of setting up and observing your own control for a given project.
- Support use of land information (and GIS application) systems for land resource planning and management.
- When designing a subdivision, comply with relevant environmentally friendly practices and regulations such as those of the local Conservation Subdivision Ordinances in the United States (see Section 4.3.3).
- Support design and installation of solar-powered continuously operating GPS/GNSS station(s) or network(s) infrastructure.
- If necessary and permissible, support use of project data to contribute to communal, regional, national, or global databases (e.g., submitting data for inclusion into the NSRS).

4.3.2 Environmental Impact

Use technology or methods with the least pollution or environmental impact. For instance, apart from being more productive and economical, use of continuously operating satellite-based GPS stations as external control to a project might reduce carbon footprint compared to driving around the perimeter of a project setting up and observing temporary control. Similarly, using satellite-generated data might be more beneficial to the environment compared to aerial mapping using low flying aircraft. These are subjective, minor examples because a number of other overriding factors may ultimately influence the method of choice for a given project situation.

FIG emphasizes that the surveyor's professional work must reflect a concern for environmental consequences and opportunities. The surveyor has an ethical duty to advise and inform upon these matters and to suggest any alternatives that may be more environmentally acceptable. Such duty shall always require:

1. An assessment of the environmental consequences of professional activities in a responsible way
2. Constant efforts to secure a recognition of environmental planning and management aspects in the fulfillment of any project, and to disseminate environmental information within the surveyors' field of expertise
3. Prompt and frank response wherever possible to public concerns on the environmental impact of projects, including, when appropriate, the stimulation of environmental actions
4. The utilization or recommendation of the engagement of additional expertise whenever the surveyor's own knowledge of particular environmental problems is insufficient to the particular task
5. The improvement of environmental standards, meticulously observing any statutory requirements on environmental issues.

4.3.3   Green Design

Certain laws and regulations are specifically meant for green design in land development (i.e., environmentally friendly design that embraces nature conservation and the use of clean energy technologies such as solar). One of those laws is the Conservation Subdivision Ordinances in the United States.

A conservation subdivision is a real estate development project in which half or more of the buildable land area is designated as undivided, permanent open space. It is a tool to realize both real estate development and conservation goals. The plan in Figure 4.2a represents a conventional development pattern in which nearly all available land is divided into house lots and streets. By contrast, conservation subdivisions (Figure 4.2b) can create not only housing and infrastructure but also permanently preserved open space. This form of development is achieved by a four-step design process that includes (1) identifying important open space features and conservation areas, (2) identifying development locations, (3) locating streets and trails, and (4) drawing in the lot lines. This approach is similar to the conventional approach, but the sequence is in the opposite order. Instead of developmental leftovers, open space becomes a valuable asset and lasting legacy. A second aspect of green design in land subdivision is the design that optimizes use of clean energy technologies such as solar. A residential subdivision design could be such that most of the buildings to be erected in the subdivision lots can make optimal use of the sun's energy throughout the year. For instance, in the Northern Hemisphere, houses built with a north–south aspect will benefit most from the sun's diurnal path. In that case, on the one hand, if a developer or building owner decides to install fixed-frame solar panels on the south-facing side of the roof, there will be maximum gain even during winter solstice when the sun's diurnal path is in the Southern Tropic. On the other hand, a similar solar panel located on the eastern

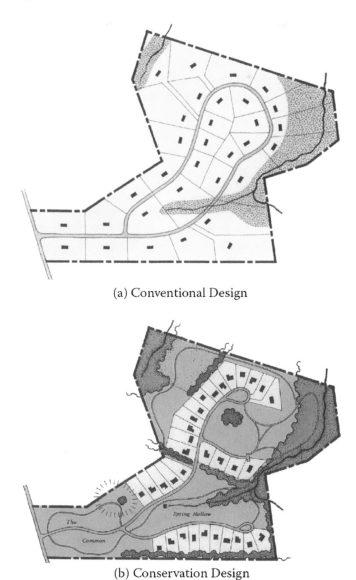

(a) Conventional Design

(b) Conservation Design

FIGURE 4.2 Subdivision design using (a) conventional development pattern, and (b) conservation development pattern.

side of the roof of a building with a west–east aspect would only benefit optimally from half of the sun's diurnal movement.

4.3.4 Case Examples

Figures 4.3–4.7 are practical examples of Conservation Subdivision design approaches (Laurien et al. 2005).

FIGURE 4.3 Solitude Point (Hamburg, Michigan). Overall density: 1 unit per acre; average lot size: 0.5 acre; open space use: wetlands, paths; percentage of open space: 50%.

FIGURE 4.4 Boulder Ridge (Green Oak, Michigan). Overall density: 1 unit per acre; average lot size: 0.5 acre; open space use: wooded, pond; percentage of open space: 50%.

FIGURE 4.5   Setter's Point (Hamburg, Michigan). Overall density: 1 unit per acre; average lot size: 0.5 acres; open space use: recreation, paths; percentage of open space: 50%.

FIGURE 4.6   Winans Woods (Hamburg, Michigan). Overall density: 1 unit per acre; average lot size: 0.5 acres; open space use: wooded, paths; percentage of open space: 50%.

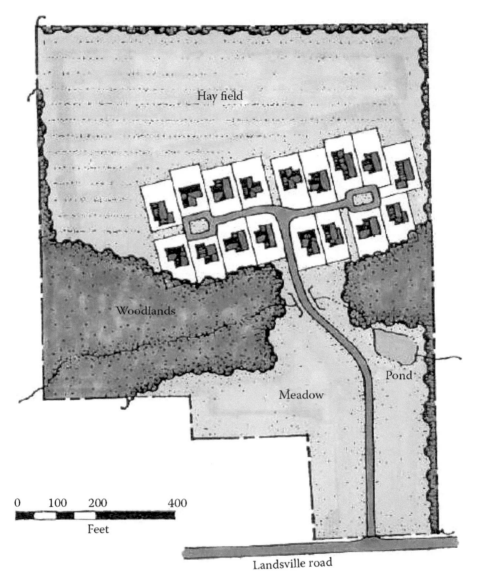

FIGURE 4.7 Canterbury (Pennsylvania). Overall density: 1 unit per acre; average lot size: 0.25 acres; open space use: hay, woodlands, pond; percentage of open space: 50%.

BIBLIOGRAPHY

Aketch, J. M. M. 2006. Land, the environment and the courts in Kenya. A background paper in Kenya Law Reports. February.

ASPRS. 2008. Guidelines for procurement of professional aerial imagery, photogrammetry, lidar and related remote sensor-based geospatial mapping services. *Photogrammetric Engineering & Remote Sensing*, November: 1286–1295.

Delaware County Regional Planning Commission. http://www.dcrpc.org/ (accessed December 12, 2009).

Dewberry, S. O. 2008. *Land Development Handbook: Planning, Engineering, and Surveying.* New York: McGraw-Hill.

Enemark, S. 2004. Building Land Information Policies. *Proceedings of Special Forum on the Building of Land Information Policies in the Americas,* Aguascalientes, Mexico, 26–27 October.

Enemark, S. 2007. Around the Globe: Surveyors and the global agenda. Professional Surveyor Magazine. September.

Enemark, S. 2009. Global Trends in Land Administration. *Proceedings of First International Conference of the Arab Union of Surveyors,* Beirut, 29 June–1 July.

Laurien, P. C., J. W. Clase and S. B. Sanders. 2005. Conservation subdivision case studies: Michigan & Pennsylvania. *Delaware County Regional Planning Commission.* Ohio: Delaware.

Morse, L. C. and D. L. Babcock. 2007. *Managing Engineering and Technology.* 4th ed. Upper Saddle River, NJ: Prentice Hall.

Natural Lands Trust Inc. http://www.natlands.org/home/default.asp (accessed December 17, 2009).

UN-Habitat. 2003. *Handbook on Best Practices, Security of Tenure and Access to Land.* Nairobi: UN-Habitat.

Wisconsin Board of Commissioners of Public Lands. *Public Records Policy.* Wisconsin: Board of Commissioners of Public Lands. bcpl.state.wi.us (accessed December 23, 2009).

EXERCISES

1. Explain the following terms with examples:
   i. Law
   ii. Rule
   iii. Regulation
   iv. Policy

2. Consider the following policy statement:

   Monumentation Policy:

   The Washington County Surveyor's Office will not accept a brass screw as an approved monument. If you request to set a type of monument other than those required by ORS, you will be required to use a Berntsen "Survey Mark (BP1 or BP2)," a MARK-IT "Mini Plug Marker (MPM-SM or MPM-PFT)," or other equivalent monument. Equivalent monuments must be pre-approved by the Washington County Surveyor.

   Carry out a research on the costs of the different monuments described in the statement.

3. Consider the following statements:

   The value of a 0.5-acre lot with 50% publicly owned open space is slightly higher than a 1-acre lot with no public open space.

   Residential lots in conservation subdivisions sell at faster rates than those in conventional developments, due to their aesthetic appeal and lack of need for private open space maintenance.

   Developers prefer conservation developments, so they can sell lots faster and don't face the cost of clearing unnecessary land of trees, shrubs, and other environmentally sensitive elements.

    Which of the following is most likely correct regarding the statements?

    a.   They all underscore the benefits of a land regulation to the environment.

    b.   They all underscore the positive social implications of a subdivision design.

    c.   They all underscore the economic benefits of a land regulation.

    d.   Only two of them underscore the economic benefits of a land regulation.

4. The government of Afghanistan has invited your company to participate in a bid for a project, which has the following components:

    a.   Update the existing 1:50,000 topographic map for Afghanistan and digitization of roads, rivers, and streams.

    b.   A set of maps for 12 major cities at a scale of 1:5,000 based on high-resolution satellite images.

    c.   Updated 1:250,000 scale land cover maps and digital database for the entire country. The information will be used for agriculture, irrigation, and sustainable natural resource planning.

    The rapid reconstruction and development of Afghanistan desperately need recent maps and topographic information. The agency responsible for such maps and information is updating very few maps due to lack of proper equipment, skills, funds, and knowledge of new technology. The first 1:50,000 scale maps for Afghanistan were produced in 1957, and a second set was produced by the former USSR government in 1975. Both map sets contain very old and outdated information, and exist only in print format. These maps hold information such as roads, rivers, streams, lakes, marshlands, settlement location, contours, place name, city location, and much more, but are static and only useful as a picture. In addition, the information in the existing maps is more than 30 years old and in the Russian language.

    Discuss some of the social and global challenges that your company may have to deal with in this project, just in case it wins the bid.

5. Figure 4.8 is a subdivision design that a developer has presented to you for ground survey. Does it comply with a Conservation Subdivision Ordinance, which requires that 50% or more of a real estate development should be public open space? Explain. (Hint: calculate the land that is not divided into lots as a percentage of the subdivision.)

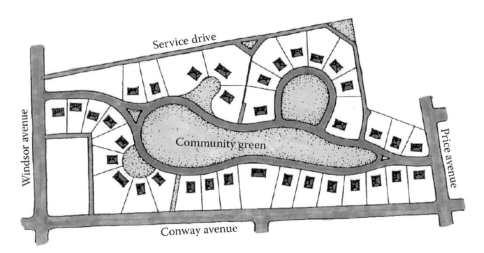

FIGURE 4.8   A subdivision design.

Part III

Planning and Design

5 Boundary Surveys

5.1 INTRODUCTION

A boundary survey is a survey to establish or reestablish the boundaries or limits of a property or easement identified by a deed of record.

> Surveyors go to the field to look for evidence of boundaries that may have been created years ago or to create new boundaries to be recovered in the future.
>
> *(Robillard et al., 2009, vii)*

The first step in performing a boundary survey is a search for evidence such as existing property corner monuments (both on the subject property and on adjacent properties), block corners, subdivision corners, and section corners. Once a sufficient amount of monumentation and other evidence (fences and other lines of occupation) are located, a series of measurements is performed. These measurements are then compared to the legal descriptions for the subject parcel and adjoining parcels. A determination of the property boundary is made, giving consideration to the accuracy of found monuments, seniority of the subject parcel, and any other evidence discovered.

After the determination of the boundary location, property monuments are placed at each angle point or change of direction, such as the beginning or end of a curve. If practical, monuments may be placed on an extension of the line. These types of monuments are called offsets, and are usually found on sidewalks or tops of curbs. Offset monuments are easier to find and to protect from damage, but may be mistaken as the actual property corner.

In the United States, when a property corner is set by a licensed surveyor, a corner record or a record of survey must be filed with the local jurisdiction. The type of filing depends on the legal description of the property being surveyed. If the legal description is a simple lot in a subdivision description, then a corner record can be filed. If the legal description is a portion of a lot or a mete and bounds description, then a record of survey must be filed.

5.2 HOW ARE BOUNDARY LINES ESTABLISHED?

Establishing boundary lines on a piece of land is not as easy as getting a measuring tape and the deed out. Land surveyors use a combination of research, science, and art to determine where the true boundaries of any given property are. Only licensed and regulated land surveyors have the skills and means to assess a property's boundary lines properly.

Most people would assume that the boundary survey starts with the measurement of the property, but in reality, the survey begins with the licensed surveyor searching for available records concerning the property such as abstracts, title opinions, title certificate, or deed. The land surveyor must investigate and take under consideration past surveys, titles, and easements. Most properties today have been created from multiple divisions of a larger piece of property. Any time a division occurred and was not properly surveyed or recorded makes the current surveyors job more challenging.

Once the surveyor has established where the boundaries historically lie, he or she will start to take measurements to see if the existing boundaries conflict with those in the historical records or past surveys. In the past, buildings, trees, or other landmarks were used as markers in property boundaries. This creates problems when trying to resurvey, as those markers may no longer exist. The surveyor can use several different tools when taking measurements, from using traditional transit and tape, or electronic distance and angle measuring equipment. Increasingly, satellite positioning equipment such as Global Positioning System (GPS) is used, although the technology is not as effective in heavily forested areas. The use of modern technology aids surveyors by gathering information quickly, but does not replace the analytical skills and assessment of the surveyor.

With the research and new measurements complete, the surveyor will then advise on any encroachments or defects in the previous description of property. The surveyor can address any specific concerns the landowner may have about these discrepancies. When all of the concerns have been addressed, the surveyor will give a professional assessment of where the true boundaries of the property lie. The property lines, as stated by a land surveyor licensed in the state where the property lies, become the legal boundaries of the property. If the new boundaries vary significantly from what the assumed boundaries were, it may be necessary for others to have a boundary survey conducted as well. It's not unusual for a land surveyor to end up performing surveys on several properties in a given area.

If there are serious legal concerns about the property, such as encroachments, easements, or a significant difference between legal property lines and an assumed one, a land surveyor will advise the client as best he or she can, but will also refer the client to a lawyer that specializes in property issues.

A good land surveyor will be able to guide the landowners through the boundary survey process and answer any question they have along the way. While there is an expense associated with getting a proper survey completed, having one done in advance of any property issues arising will result in greater peace of mind regarding property ownership.

5.3 BOUNDARY TYPES AND BOUNDARY MARKERS

From Biblical times when the death penalty was assessed for destroying corners, to the colonial days of George Washington who was licensed as a land surveyor by William and Mary College of Virginia, and through the years to the present, natural objects (i.e., trees, rivers, rock outcrops, etc.) and man-made objects (i.e., fences, wooden posts, iron, steel or concrete markers, etc.) have been used to identify land parcel boundaries.

—Ghilani and Wolf, 2008, 621

A boundary is a line of demarcation between adjoining land parcels. The land parcels may be of the same or of different ownership but distinguished at some time in the history of their descent by separate legal descriptions.

Boundary types can be described according to their primary purpose. For example, *subdivision* and *lot* boundaries and *easement* and *right-of-way* boundaries are the most common types in real estate dealings and in land development activities. *Riparian* and *littoral* boundaries are the more complex boundaries formed by the waters of a river, lake, sea, or another body of water. The general rule is that riparian boundaries shift with changes due to accretion or erosion but retain their original location if brought about by avulsion or by artificial causes. A littoral boundary is the boundary of the shore, especially a seashore subject to tidal inundation where submerged lands meet uplands—more specifically defined by a tidal datum determined by observing long-term tide levels. Political or administrative boundaries such as *national* and *state* boundaries between countries and states are established by treaties made by the concerned sovereign powers.

A land boundary may be marked on the ground by material monuments placed for the purpose, such as fences, hedges, ditches, roads, and other service structures along the line. They may also be defined by (1) astronomically described points and lines, (2) coordinates on a survey system whose position on the ground is witnessed by material monuments that are established without reference to the boundary line, (3) reference to adjoining present or previous owners, and (4) by various other methods.

5.4 BOUNDARY SURVEY DESIGN AND PROCEDURES

5.4.1 INFORMATION GATHERING

Reproducing previously established boundaries is a matter of determining the intent of the original survey or other written documents first used to establish them. Research is the key for a successful boundary retracement. That research includes gathering evidence, interpreting that evidence in light of conditions at the time of the creation of the boundary, and, finally, re-creating the location of the boundary as it exists today. Consideration of written evidence is of utmost importance, but the physical evidence and oral testimony related to interpreting the written evidence are also crucial to completing the research.

Research for boundary surveys begins by gathering property descriptions, tax maps, roadway plans, easements, and other related information. Important sources of such information include local courthouses, utility companies, and state and federal agencies. Title insurance companies can provide information such as deeds, easements, other record title information, and previous surveys of the subject property or adjoining properties. However, title insurance companies often do not research far back enough in time to return to the original survey.

The individual pieces of information gathered from research (such as maps, deeds, and patents) can be plotted together to form a logical composite map. The composite map provides pieces of a jigsaw puzzle for important clues to the interpretation of land ownership and boundary location.

5.4.1.1 Encroachments and Gaps

Visible evidence of encroachments of any kind must be located and their history documented if possible. An encroachment is the gradual, stealthy, illegal acquisition of property—the act of trespass upon the domain of another or the commencement of a gradual taking of possession or rights of another. They may take the form of buildings, walls, fences, or other structures representing occupation or possession, which extends onto another's title rights—a building or part of a building that intrudes upon or invades a highway or a sidewalk, locating factories in a residential district, and so forth.

Encroachments may lead to claims of unwritten rights such as adverse possession or acquiescence when the party whose land is encroached upon allows the encroachment to go unchallenged. Such an implied consent may lead to loss of title in some doctrines.

If the project involves several smaller tracts joined together to form a larger, more developable tract (as land contemplated for development), it is important to confirm the continuity of land parcels and to identify any gaps that may exist between them. This is critical because the record title to such a gap may lie in the hands of some former owner rather than in the hands of the owner of the parcels being surveyed.

5.4.1.2 Research of Land Records

The actual process of conducting research varies widely both nationally and internationally. All information from public and private records that may be of benefit to performing the boundary survey must be sought out. The information is ideally gathered and evaluated prior to commencement of fieldwork. If not, additional trips to the field may be required as new evidence from records dictate further field investigation.

Research information comes from a variety of sources. For example, in the following paragraphs let us consider briefly the research process of land records in the United States.

Typically, a visit to the county assessor or auditor will help in the search for the record descriptions of the subject property and its adjoining properties. Once the names and property transfer dates are determined, a copy of the actual transfer document can normally be found in the county registry of deeds, county recorder's office, or clerk of the court.

All states allow for the recording of a deed when the transfer of real property takes place, and almost all such transfers by deed are recorded. These records should be used to check previous transfers of the subject property and the adjacent property. Boundary location intentions are obtained from the information at the time the lines and corners were created. Comparison of the deeds of the subject property with those of adjoining property should be made to determine if any conflicting calls or descriptions exist.

5.4.1.3 Title Search

The type or method of record keeping varies from county to county and state to state, and may include methods such as transfer books, grantee/grantor index books, deed record books, and online digital format using GIS systems. To use the information, you must be familiar with the system used to index the thousands of documents, usually chronologically by date of recording. In public land states of the United States, surveyors often rely on reports, public records, field notes of the original surveyor, and reports prepared by the title companies providing title insurance to the lender and/or buyer. However, information provided by title companies does not eliminate the liability of the surveyor to conduct an independent deed search or establish a chain of title for the property.

In the grantee/grantor index system, for example, the process of checking the title begins with the name of the present owner in the grantee index. The search is to determine intent of the parties at the time of the original survey or, if there was no original survey, the intent of the parties based on the original writings. The search is done back in time to establish a chain of title for the property. By reviewing each conveyance, it can be determined whether any owner in the chain of title has encumbered or impaired title to the land with, for example, easements, other sales, mortgages, or liens. In a chain of title, links (connections or transfer of property between consecutive owners) are also carefully reviewed to establish the property transfer. These links are not always in the form of a deed but can be a will, the records of intestate estate (an estate left by a deceased without will), or a court order.

Surveyors are particularly concerned with the boundary description in each conveyance.

5.4.1.4 Adjoining Property

As part of a normal standard of care, the deed research should include adjoining properties as well. Adjacent deeds are checked using the most recent descriptions. Boundary lines of the subject property are compared with those of the adjoining property to determine if there are any discrepancies. If any problems exist, the deed for the adjoining property might also have to be checked back in time to determine the intention of the original parties when the line was created.

When the calls on the same lines in adjoining deeds do not agree, it may be necessary to establish senior rights for the line in question. *Senior rights* are the rights gained by virtue of being the first parcel created out of a tract.

5.4.2 ANALYSIS OF INFORMATION

Having assembled all necessary records about a property, the work of examining those records for determining boundaries can begin. It is the surveyor's job to interpret the intent of the descriptions and to place the described boundary lines and corners on the ground according to accepted principles of law. Much emphasis is placed on the intention of the parties at the time a monument or boundary originates. It is therefore important to know how to find that intent in a manner that will be upheld in a court of law.

A composite map of the subject property and adjoining parcels is prepared using all appropriate information gathered during the research phase. Each map or description is prepared at the same scale as (and in reference to) the subject property. The composite map will include bearings and distances and calls for every line and corner represented. This becomes the guide for determining where evidence of the boundaries might be found during field surveys, and any potential conflicts between the subject and adjoining properties.

5.4.2.1 Monuments

The next step in the decision process is the use and application of existing monumentation on the ground to determine the physical location of boundary lines. For this step, it is important to understand the difference between existing monuments (corners), obliterated monuments (corners), and lost monuments (corners). The following definitions are taken from the *Definitions of Surveying and Associated Terms*, 2005:

> An *existing corner* is a corner whose position can be identified by verifying the evidence of the monument, or its accessories, by reference to the description in the field notes, or located by an acceptable supplemental survey record, some physical evidence or testimony.
>
> An *obliterated corner* is a corner at which there are no remaining traces of the monument or its accessories, but whose position may be recovered beyond reasonable doubt, by the acts and testimony of the interested landowners or based on other supplemental evidence.
>
> A *lost corner* is a previously established corner whose position cannot be recovered beyond reasonable doubt. The location of a lost corner can be restored only by reference to one or more of other independent corners.

5.4.2.2 Discrepancies

Consistency of calls for subject property monuments with those of the adjoining properties must be checked. Any differences should be noted and reconciled if possible.

Discrepancies in line bearings may result from variation in magnetic north between the time the description was constructed and the present reference to north. Variations in distances can also exist. Both human error and technology can be a factor in discrepancies between calls in different documents.

Bearings are often copied from older deeds. They are important for the proper preparation of the composite map. The date of the survey that produced the bearing must be known so that corrections for magnetic declination can be properly made. If different documents gathered from the title search have different references for the north, they can be adjusted to a common north reference system such as True North or Grid North.

5.4.2.3 Fieldwork Preparation

Based on the analysis of existing boundary information, a boundary survey map such as shown in Figure 5.1 is prepared for use in conducting the fieldwork.

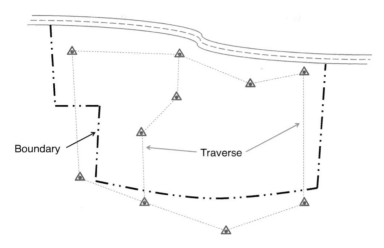

FIGURE 5.1 Boundary survey field map.

The map should include point and line information gathered from the various deeds and documents (of both the subject property and adjoiners). A horizontal traverse may be designed to control the boundary survey, although other methods of choice such as GPS control surveys also exist (see Chapter 6).

5.4.3 BOUNDARY SURVEY

In applying the information to the boundary survey, appropriate standards should be consulted (see Chapter 2). Boundary surveys for land development differ from other types of boundary surveys. Standards also vary significantly from state to state and country to country. All standards outline minimum requirements, and surveys must meet these requirements in addition to any site- or client-specific requirements.

5.4.3.1 Fieldwork

The fieldwork would begin with the reconnaissance to locate corners or other calls referred to in the gathered descriptions. The reconnaissance also serves as a basis for planning subsequent survey and for carrying out pertinent administrative procedures such as notifying adjoining landowners of impending survey. The field survey will proceed more smoothly if the cooperation of adjoining landowners is secured.

Having established the standards and carried out a reconnaissance, a traverse of the subject property maybe conducted using conventional equipment and procedures such as measuring angles and distances with total station, or the use of GPS receivers to establish locations using satellites may be employed. A combination of procedures and equipment may be necessary in some cases.

Stations in the survey should be located advantageously and referenced for future recovery (Figure 5.2)—keeping in mind that they will be used to locate features on the subject property and possibly for other future uses unrelated to the current survey (such as topographical surveys and construction stakeout and control surveys).

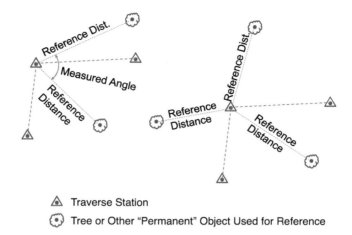

FIGURE 5.2   Methods to reference traverse stations (traverse ties).

Information to be located in relation to the boundary traverse should be properly tied to the traverse line (see, for example, Figure 5.3). This is important when locating markers and monuments, which form an integral part of the boundary resolution.

All ties and references should be accurately described and recorded in the field notes. Redundancy in measurements and correct procedures eliminate blunders and keep human errors in check.

The lines of the boundary traverse should follow the record lines when possible because the traverse serves as the basis for locating evidence that determines the position of the actual property line.

Location of features such as drainage ditches, streams, creeks, rivers, tree lines, fences, walls, right-of-way lines, access roads on or across property, and objects used

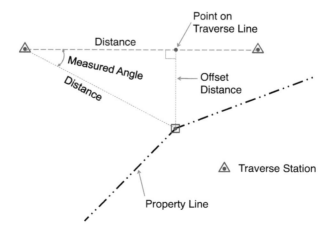

FIGURE 5.3   A boundary point tied to a traverse using two references.

to mark corners and points on the boundary line should be determined and recorded in the survey field notes. Locate these features if they are mentioned as a boundary or if they cross a boundary line.

If applicable, existing and potential future easements should be identified and located at the time of the conduct of the boundary survey because they may impact future boundary definition or land development plans. This can be done by referencing service lines of different types that adjoin or cross the subject property.

In the case of a boundary survey for land development, it is important that any evidence of possession discovered during the survey is investigated and recorded in the field notes. Such evidence may include, for example, the cutting of timber, farming, or construction.

5.4.3.2 Office Work

Once the fieldwork phase is completed, the information is processed and compiled into a presentable format. A determination of the boundary locations can then begin. All field data should be checked. Field information on lines and points should be compared with lines and points called for in the record documents. Aerial photographs and the composite map gathered earlier can aid in this process as well.

The computations should include checks for mathematical integrity (such as traverse closures) and also analyses to check if the pertinent standards have been met. The integrity of the survey control data ensures that all points included in the survey (such as fences, monuments, and references) have coordinate values that are legally traceable and defensible. All these points are plotted in their respective positions relative to each other.

Using accepted guidelines, various pieces of evidence are given their appropriate weight and the locations of missing corners are computed. The relationship of these points to the nearest traverse line is established so that a field search can be made to determine the boundary.

The survey computations form a basis for boundary determination by walking all lines of the survey searching for points called for but not found and also looking for features such as fences, utility lines, roads, encroachments, streams, or any other objects inadvertently omitted during the fieldwork. Nothing affecting the location of the boundary should be overlooked.

5.5 LEGAL CONSIDERATIONS IN BOUNDARY DETERMINATION

In cases where relocated corners do not agree with all calls of original documents or notes, the rules for considering physical evidence have a particular order of priority as follows: (1) natural monuments, (2) artificial monuments, (3) calls for adjoiners, (4) courses and distances, and (5) acreage. When the inconsistency is between course and distance, the distance is generally considered more reliable, but not necessarily more legal.

All information gathered from the land records search and the fieldwork must be determined in making the final determination of a boundary. The laws governing property boundary surveys and real property also play a role in the decision process and in the weighting of evidence to support claims. Final boundary determination rests with the courts, if the boundary problems are not resolved by the surveys.

5.5.1 CONFLICTING TITLE ELEMENTS

In making decisions based on discovered evidence, familiarity with the order of importance of conflicting title elements is important. These elements are briefly discussed in the following paragraphs:

Boundaries by possession: The first element is the right of possession. These are the rights not stated in writing but can become rights in fee. These unwritten rights often involve adverse possession, although there are a number of other types of unwritten rights such as acquiescence, estoppel, and oral agreement. Such possession are often initially determined by a land survey, and then ultimately by a court of law. A land surveyor cannot make a determination of the legality of the possession.

Boundaries by senior rights: Senior rights are the next element in order of importance. They are defined as the rights in a parcel of land, created in sequence with a lapse of time between them (Genovese, 2005, 222). A person conveying part of his or her land to another (senior) person cannot, at a later date, convey the same land to yet another (junior) person. A buyer (senior) has a right to all land called for in a deed.

Boundaries by written intentions: The next element in consideration is the written intentions of the parties involved. These intentions are those expressed by the parties to a conveyance and put in writing in the document that brings about that conveyance. Such written intentions may themselves include a call for a survey of the boundaries.

Calls for monuments: As previously stated, natural monuments take precedence over artificial monuments. In some cases, artificial monuments can be so well identified that they become of equal importance. Monuments of record (calls for adjoiners) are often considered a third type of monument.

Distances, bearings, areas, coordinates: After consideration of monuments, elements such as distances and bearings as called for in the property descriptions are next in the order of importance. Areas do not have a high priority except in cases of wills. Coordinates are also given consideration, but they could be subject to mistakes especially if they are computed from distances and bearings. Conveyancing parties rarely understand the relationship of coordinates to a field location, so they might not accurately represent the intentions of the parties.

Boundaries by agreement: A boundary can also be established by mutual agreement between the landowners. Confirming such agreement is dependent on finding where the mutually agreed boundary is located. Courts often require that certain factors be present before boundaries can be established based on such agreements. For instance, the agreements must set out a specific line as the boundary; the parties in the agreement must occupy the adjoining lands; and the agreement must be recognized for a considerable time.

Boundaries by estoppel: Boundaries by estoppel develop when the true owner knowingly misrepresents the boundary causing a neighbor to rely on the misrepresentation. In such cases, the boundary becomes as represented. In the *Definitions of Surveying and Associated Terms* (2005), estoppel is defined as "stopping a person from asserting a claim by reason of his own previous representations which refute his new claim. The new claim may in fact be true, however, he may be prevented from exerting that claim by *estoppel*." For example, a property owner who knows that an adjoiner is making improvements along a line, which they have incorrectly believed to be the true boundary, may later be estopped from claiming the true boundary line (Figure 5.4).

BIBLIOGRAPHY

Dewberry, S. O. 2008. *Land Development Handbook: Planning, Engineering, and Surveying.* New York: McGraw-Hill.

Douglas County, Oregon. http://www.co.douglas.or.us/survmaps/Survey/M149/M149-65.pdf (accessed January 27, 2010).

Genovese, I. (ed.). 2005. *Definitions of Surveying and Associated Terms.* Gaithersburg, MD: American Congress on Surveying and Mapping.

Ghilani, C. D. and P. R. Wolf. 2008. *Elementary Surveying: An Introduction to Geomatics.* 12th ed. Upper Saddle River, NJ: Prentice Hall.

Robillard, W. G., D. A. Wilson and C. M. Brown. 2009. *Brown's Boundary Control and Legal Principles.* 6th ed. New York: Wiley.

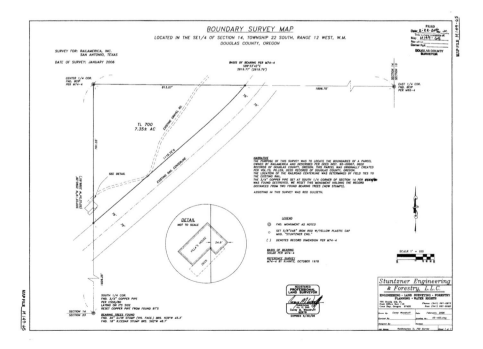

FIGURE 5.4 Sample boundary survey map. (Courtesy of Douglas County, Oregon.)

EXERCISES

1. Discuss the following aspects of boundary survey design and procedures:
   a. Information gathering
   b. Information analysis
   c. Application of gathered information to the boundary survey
   d. Legal considerations in boundary determination
2. Case Study: Discuss the boundary survey procedures in your local area or county. In your project, include the aspects of standards and regulations that pertain to boundary surveys, record or title search methods, and any other matters that pertain to a local boundary survey. What specific challenges are unique to your location?

6 Control Surveys

6.1 GENERAL CONSIDERATIONS

6.1.1 PROJECT SCOPE AND REQUIREMENTS

A control survey design should take into account: (1) the number and physical location of project points (i.e., distribution and geometry), (2) the standard errors you expect to achieve from field measurements, and (3) the number and types of observations to be measured (repeat measurements and instrument types). With these variables, especially the first two, a presurvey analysis may be carried out (e.g., by using least squares propagation) to assess the best design alternative (Section 6.4.4).

6.1.1.1 Number and Physical Location of Project Points

If the project does not specify how many new project points are required and their respective distribution within the project area, then local standards and regulations could form the basis for the network design. For instance, in the United States, a first-order triangulation would require a station spacing of not less than 15 km (FGCC, 1984). Conversely, the accuracy standard and size of the project area will determine the number and distribution of control points. Other factors to consider include, for example, the physical access and interstation visibility or sky visibility, depending on whether traditional or GNSS methods will be used.

6.1.1.2 The Layout of Project Area

An initial visit to the project site is not always possible. However, online browsing makes it possible to virtually assess the site from another location, even thousands of miles away on another continent. The influence of site topography on project design and logistics is a primary consideration. For example, site accessibility and transportation from station to station must be given careful thought. Some areas may only be accessible by helicopter or other special vehicles. Roads may be excellent in one part of the project and poor in another. The general density of vegetation and buildings, or the primary economic activity in an area may lead to general questions of overhead obstruction or multipath (i.e., site-specific characteristics). Information of land ownership is important for logistics such as obtaining permission to cross property.

Maps and imagery are particularly valuable for preparing a GNSS survey design. Depending on the scope of the survey, various scales and types of maps can be useful. For example, a GNSS survey plan may begin with the plotting of all potential control and project points on a map of the area.

DOI: 10.1201/9781003297147-9

6.1.1.3   Accuracy and Datum of Control

The accuracy of a control point is determined by the accuracy of the survey and the quality of data adjustment. It is a function of the equipment, methods, and procedures used, as well as any external control applied in the project. If the project is tied to existing external control, we usually speak of absolute positions of points, in a given coordinate system (such as WGS84, NAD83, State Plane, and UTM). Alternatively, we speak of relative accuracy and the internal relationships between project points.

If external control is applied in the project, it is based upon hierarchical classification of accuracy whereby new control is referenced to previously existing control of higher order. Many standards exist for geodetic control, depending on the type of survey work to be accomplished and the methods to be used (see Chapter 2).

- Primary or First-Order control is used to establish geodetic points and to determine the size, shape, and movements of the earth.
- Secondary or Second-Order Class I control is used for network densification in urban areas and for precise engineering projects.
- Supplemental or Second-Order Class II and Third-Order control is used for network densification in nonurban areas and for surveying and mapping projects.

6.2   GNSS CONTROL SURVEYS

The benefits of using GNSS in support of conventional methods have been noted in many textbooks and much of the literature (e.g., Ogaja 2011; Ogaja 2022; Rizos 1997; Van Sickle 2007). *Inclement weather does not disrupt GNSS observations*, and *a lack of intervisibility between stations is of no concern whatsoever*. The following sections will present important considerations in the planning and design of GNSS control surveys.

6.2.1   INDIVIDUAL SITE CONSIDERATIONS

Environmental factors, such as physical objects causing *obstructions* of GNSS signals and *multipath*, cause errors in GNSS data collection.

Generally, GNSS signal reception is better in an open field than under a tree canopy or in a natural or urban canyon. Therefore, try to avoid obstructions caused by buildings or vegetation during data collection. If you cannot avoid them, plan to collect data at these locations when there are a maximum number of satellites in the sky. Greater sky visibility at the antenna location provides more accurate data. Most commercial GNSS software provide planning utility for checking satellite signal availability at a planned GNSS site. For example, you can check the potential dilution of precision (DOP) of the site on the basis of a visibility plot prepared from an actual site visit (Figure 6.1).

DOP is a measure of the quality of GNSS positions based on the geometry of the satellites used to compute the positions. When satellites are widely spaced relative to each other, the DOP value is lower, and position accuracy is greater. When satellites

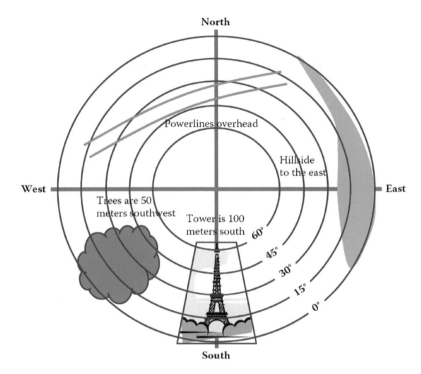

FIGURE 6.1 Site visibility obstruction diagram.

are close together in the sky, the DOP is higher and GNSS positions may contain a greater level of error. There are five different types of DOPs that can be defined, but the geometric DOP (GDOP) is the most commonly used. A mathematical definition of GDOP and its potential application in the GNSS survey design are presented in Section 6.4.4.

Multipath occurs when the GNSS signal is reflected off an object before it reaches the GNSS antenna (Figure 6.2). Multipath error occurs without warning. The error can be minor or can result in several meters of accuracy degradation. High multipath surfaces include urban canyons, dense foliage, and generally large tall structures. Currently, there is no way to prevent multipath from occurring except to *minimize its effects through careful planning and site selection.*

A high multipath location such as in urban canyon may also imply limited satellite visibility, and hence high DOP values. Low multipath environments such as open sky areas are likely to offer good (low) DOP values. Overall, it is important to check the potential DOPs of a planned GNSS site on the basis of a visibility plot prepared from an actual site visit (Figure 6.1).

6.2.2 CONTINUOUSLY OPERATING GNSS NETWORKS

Several organizations such as the National Geodetic Survey (NGS; www.ngs.noaa. gov) and International GNSS Services (IGS) provide continuous tracking data,

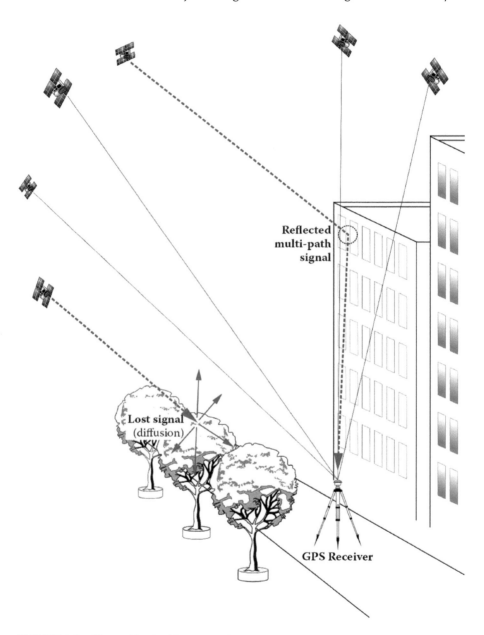

FIGURE 6.2   Site multipath. (From *GPS for Land Surveyors*, Boca Raton, FL: CRC Press, 3rd ed., 2008, Figure 2.4, 44. Used by permission.)

available online, that can be downloaded for everyday GNSS applications (Ogaja, 2022). The online GNSS data, originating from continuously operating reference stations (CORS), can serve as external control data when carrying out a GNSS project.

Currently, when designing a survey plan by plotting all potential control and project points on an existing map, CORS are not available yet on such maps. However, in early 2009 the NGS launched a utility (named NGSCS) that can aid surveyors in project planning, in terms of locating existing survey control in the vicinity of a project location. The NGSCS is an Internet-based interface for visualizing the NGS control stations in Google Earth (earth.google.com). A user intending to apply this utility for project planning is first required to download and install both the Google Earth and a NGSCS kml file, free of charge. Google Earth uses geodetic coordinates in the WGS84 datum, although as a precaution *the coordinates, elevations, distances, and measurements provided by Google are approximations only.*

Once installed, a NGSCS Radial Search creates a Google Earth network link that plots the approximate location of NGS control stations within a specified distance of the view center (Figure 6.3). The view center is the point of interest within a project area. Station markers indicate the type of control station as follows: H (orange) = horizontal only, V (blue) = vertical only, and B (pink) = both horizontal and vertical. If the search does not yield any GNSS CORS within the programmed radius of a project location, an alternative is to locate the CORS data directly from the NGS or other related Web site.

6.2.2.1 GNSS Processing Using Online Services

The greatest benefit of GNSS CORS to the surveying community is the possibility to execute a control survey project *even with a single GNSS receiver.* A single GNSS operator can occupy all the project points one at a time and still be able to tie all the points to an external CORS network at the data processing stage. The existence of *online CORS-based GNSS processing services* that are freely available and remotely accessible from any location is an important element of this possibility.

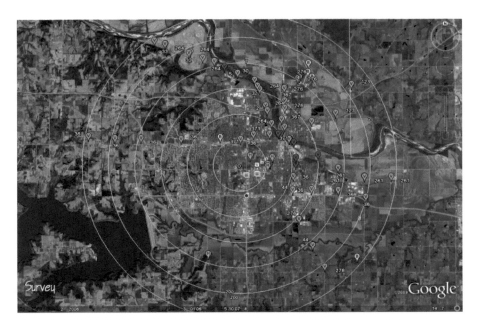

FIGURE 6.3 A Google Earth image of NGS control stations within 6 miles of a project location. Each ring represents a radial distance of 1 mile.

The online GNSS processing services allow remote processing of GNSS data files submitted via the Internet. This reduces project costs significantly, especially for projects that require external control. During remote processing, CORS data are either applied as external control or to estimate and remove some of the GNSS errors that degrade the position coordinates. Here are some of the free online GNSS services as they existed in 2022:

- *OPUS (Online Positioning User Service) (www.ngs.noaa.gov/OPUS):* A NGS utility that allows for the processing of GNSS data for static surveying methods. It computes an accurate position for submitted data file and ties the coordinate results to any three closest NGS CORS.
- *CSRS-PPP (https://webapp.csrs-scrs.nrcan-rncan.gc.ca/geod/tools-outils/ ppp.php):* The CSRS online database allows users direct access to the primary horizontal and vertical control networks archived on the *Canadian Geodetic Information System (GIS)* and the *CSRS—Precise Point Positioning (PPP) online GNSS processing.*
- *AUSPOS (https://gnss.ga.gov.au/auspos):* AUSPOS provides users with the facility to submit dual-frequency geodetic-quality RINEX data observed in a *static* mode, to a GNSS processing system and receive rapid turn-around *Geocentric Datum of Australia* and *International Terrestrial Reference Frame* coordinates. This service takes advantage of both the IGS Stations Network and the IGS product range, and works with data collected anywhere on earth.
- *SCOUT (Scripps Coordinate Update Tool):* SCOUT processes RINEX files to calculate precise coordinates and also allows users to obtain input files for their processing software. It gives the option to select up to four reference CORS to be used in the processing. The Scripps Orbit and Permanent Array Center (sopac.ucsd.edu) is located at the Cecil H. and Ida M. Green Institute of Geophysics and Planetary Physics, Scripps Institution of Oceanography, University of California, San Diego in La Jolla, California.
- *GAPS (GPS Analysis and Positioning Software):* GAPS (from the University of New Brunswick) processes submitted RINEX observation files in either static or kinematic modes (gaps.gge.unb.ca). It specifically uses the UNB3m neutral atmospheric delay model, and solutions can be optionally constrained (weighted) to a priori coordinates (either specified or from RINEX file). Results are returned via e-mail.

Although these services are available for use worldwide, the mode of application may be limited for project locations where CORS network sites are few and far between. In reality, many parts of the world still have no (or very little) reference control networks established, limiting the use of online services. Nevertheless, some of the services (such as CSRS-PPP, AUSPOS, and SCOUT) would still suffice for those locations, especially for applications where centimeter-level accuracy is sufficient.

6.3   TYPICAL WORKFLOW OF A GNSS PROJECT

As with boundary surveys, every GNSS project is different. It is possible to list typical GNSS project components. While not all of these components are part of the
planning process, they are listed here to provide a common frame of reference for
further discussion.

1. An entity or a client determines that a GNSS project is necessary.
2. An RFP is sent to potential GNSS consultants.
3. Proposals are prepared by interested consultants.
4. A winning proposal is selected.
5. Station selection and recon begins.
6. Collect all necessary information about available control.
7. Prepare equipment inventory.
8. Prepare field data sheets.
9. Conduct necessary training for field and office persons.
10. Manage personnel (assign crew tasks).
11. Determine optimum observation times, satellite availabilities, obstructions,
    PDOP, VDOP, GDOP, etc.
12. Determine when troublesome monuments must be visited.
13. Coordinate site access if required.
14. Supervise, coordinate, communicate, and ensure safety.
15. Develop detailed plans and schedules.
16. Work up an observation (or installation) plan.
17. Conduct the fieldwork (or installation).
18. Process the data and adjust the network.
19. Prepare a final report and submit it to the client.

A brief discussion of these items is essential. The list appears to have a private company slant. The reader working for a public agency or utility company should just
read wording into the statement that simulates their particular situation. A city engineer might come into the office of the city surveyor and say, "We need to establish
geodetic control throughout the northern half of the city at a one mile spacing so prepare a proposal for me outlining how you would do the job." While this may not be
a competitive bid situation, the parallels and differences should be obvious. Despite
the particular employment situation the prospective GNSS project manager enjoys,
almost everyone should be quite familiar with items 1–4.

6.3.1   STATION RECON

Send out crew persons that will also be conducting the observations later. In this way
they are becoming familiar with the project area and learning how to best use the
transportation network and how best to safely approach, occupy, and leave project
points. If potential sky obstructions are present at a particular station, a skyview
plot should be carefully prepared. Monument and local perspective photographs,

approximate geodetic coordinates (and perhaps heights), station occupation guidance, and proposed safety measures may also be appropriate.

6.3.2 EXISTING CONTROL

Gather as much information as possible about nearby existing geodetic monuments that might be used to control the final GNSS network. Station descriptions, geodetic, state plane, or other appropriate coordinate information and if possible the accuracy characteristics of the monuments in question are needed. Look for national, state, regional, and locally available horizontal and vertical control that may be of help (see, e.g., Section 6.2.2).

6.3.3 EQUIPMENT INVENTORY

It is imperative to have a comprehensive field equipment inventory sheet. It will be a major embarrassment if a field crew gets 83 miles away from the office, with 3 minutes before the start of session 1 only to then discover that they forgot the GNSS antenna. Specific equipment and support material will vary from one application to another. Use common sense and experience to determine specific requirements for your particular application. Provide a separate inventory data sheet for each field crew and insist they actually check it over each day before heading to the field. Inventory items may include the following:

1. GNSS receiver
2. Antenna w/cables
3. Power supply (fully charged) w/cables
4. Backup power supply
5. Tripod/tribrach/tribrach adaptor
6. Tape, ruler, rod, or built-in antenna height measuring device
7. Communication device
8. Meteorological equipment (temp/pressure/humidity)
9. Data logging capability
10. Instruction manuals/books
11. Field data sheets w/pen or pencil
12. Calculator
13. Observation plan (by session)
14. Preliminary station coordinates
15. Lighting devices for night operations
16. First aid kit
17. Other.

6.3.4 FIELD DATA SHEETS

Most field GNSS data can be entered into the receiver in the field prior to and during a particular observation. It is still wise to have each crew prepare a written log of GNSS field observation activities. Maintaining this kind of field crew discipline may provide office personnel with the additional evidence necessary to resolve ambiguous results when the data are processed. Field data sheets may be specifically tailored to meet specific project goals. Information crucial to the proper processing of the data must be entered here. Not all items are mandatory for a given application. A listing of potentially necessary items include the following:

1. Date of observations
2. Station identification
3. Session identification
4. Receiver/antenna serial numbers
5. Crew person name(s)
6. Antenna height measurement(s)
7. Station diagram and equipment deployment
8. Obstruction diagram(s)
9. Actual start and stop times
10. Weather
11. Comments (especially problems or difficulties encountered)
12. Other.

6.3.5 TRAINING AND MANAGEMENT

Ensure that all project personnel are properly trained for the task at hand. Train in-house if possible or get it as part of the GNSS purchase or lease agreement. Pay for special training if necessary. Maintain strong management interaction with field and office personnel throughout the entire conduct of the project. Know their strengths and weaknesses. Cross-train as a matter of precaution (you never know when a critical employee might leave, get sick, or retire).

6.3.6 EVALUATE SITE CHARACTERISTICS

Employ vendor planning software (or other available planning software) and a recently acquired almanac (satellite ephemeris message) to investigate the health and geometry of solution for planned sites. Using site obstructions, prepare geometry plots for the proposed observation period.

6.3.7 DETERMINE WHEN DIFFICULT MONUMENTS MUST BE VISITED

Decide when to observe stations having limited access or limited sky visibility.

6.3.8 COORDINATE AND SUPERVISE

Coordinate site access when necessary. This will facilitate field crew efforts and minimize operational problems. If occupying a monument in a heavy traffic location it would be wise to contact the appropriate officials ahead of time. Support the crews during data collection and processing. Ensure that a safe working environment exists. One unfortunate accident could endanger a person and/or threaten an entire organization. The development of optimal observation schedules will be the topic of Section 6.5.3.

6.3.9 Work Up Detailed Observation Plans

A detailed observation plan allows all field crew members know who does what and where they do it. A detailed observation plan gives your field personnel a fighting chance to get the task accomplished. In the early days of Global Positioning System (GPS), long static observations were required. More recently, better technology has led to far shorter observation periods. Now, fast static techniques used for moderately spaced monumentation (1–20 km) may require that planned between station move time be much longer than observation time.

The detailed plan may include a network diagram with clearly annotated session times and the crew responsible. Such planning should be based on DOP values obtained from station recons.

6.3.10 Control Network Optimization

The control network optimization can be considered at three different levels, namely, optimizing the number of receivers, optimizing network geometry, and optimizing network observational design.

In the first design, simulations can be run using each possible array of receivers (e.g., 2–8) and a wide array of project sizes (e.g., containing between 1 and 52 baseline vectors). The optimum number of receivers can then be determined for the type of application such that the optimum number of receivers, used efficiently according to given criteria, would be most cost-effective.

In the second design, different geometries, such as *radial survey* of each network point from the closest control, simple *traverse* linking each point in succession, or a *chain network*, can be considered on the basis of the number of available receivers, observations times, network redundancies, and achievable accuracies. A geometry with shorter observation periods combined with more network redundancy and best overall combination of positional reliability and decreased cost for the survey would be the preferred geometry.

In the third design, plan the observational scheme so that as many points as possible are visited by every crew member. Remember that you only get $(r-1)$ independent baseline vectors during each session, where r is the number of simultaneously operating receivers (see also Section 6.5.2). Some techniques that have been suggested in the past include the following:

1. Begin observing check baselines between pairs of existing control points first so that the field crew members will have time to learn or remember how to properly collect the data. *This step may not apply when using CORS data for control.*
2. Start each field day by observing baselines far from the office and work back toward the project headquarters minimizing long drives for tired crews.
3. Move as few units as possible between sessions.
4. Move all units that must be moved the shortest distance possible.
5. When a long move is required, ensure that this is the optimum time required to move more than the minimum number of units.

6. Try to close as many observational loops each day as possible. If data processing keeps up with the data collection, this allows the office work to check loop closures. If the data are good, confidence is gained sooner. If the data are bad, it is detected sooner, allowing more time for observational plans to be modified.

Do not be afraid to try innovative observational schemes and don't hesitate to erase your early efforts if necessary and start again.

6.4   DESIGNING A GNSS SURVEY NETWORK

Client and accuracy requirements will influence the choice of method and field procedures to use when designing a GNSS survey network. Technology and procedures should be appropriate for the task, hence the *need to know all GNSS methods and procedures, and their strengths and weaknesses.* The general checklist items to consider include, for example, the number of stations, where to locate them, their distributions, and connections; what the standards and specifications say (e.g., how to connect to datums, equipment to be used, and field procedures to be followed); planning aids such as site visibility obstruction diagrams and a least squares simulation program; and logistical, social, and economic considerations (e.g., site access and conditions, permissions, and cost alternatives).

6.4.1   STANDARDS AND DESIGN CRITERIA

Since the advent of GPS, it has been necessary to redefine standards. The Federal Geodetic Control Subcommittee (FGCS, 1989) proposed provisional accuracy standards for GPS-relative positioning techniques to be used alongside existing standards. The older standards of *first-*, *second-*, and *third-order* are now classified under group C in the new scheme.

Although new standards accommodate control survey by static GNSS methods, not every GNSS survey job demands the highest achievable accuracy neither does every GNSS survey demand an elaborate design. For example, in open areas that are generally free of overhead obstructions, group C accuracy may be possible without a prior design of any significance. In some situations, a crew of two, or even one surveyor may carry out a GNSS survey from start to finish.

The FGCS's new standards of B, A, and AA are, respectively, 10, 100, and 1,000 times more accurate than the old first order. However, the attainment of these accuracies does not require corresponding 10-, 100-, and 1,000-fold increases in equipment, training, personnel, or effort. They are well within the reach of private GNSS surveyors both economically and technically.

A GNSS survey typically requires the occupation of *new* stations and stations whose coordinates are already known, in either a GNSS datum or the local geodetic datum. Use of survey datums is required by GPS standards and specifications: (1) for determination of local transformation parameters—between GPS datum and local datum, (2) for quality control purposes, (3) to connect new GNSS stations into surrounding geodetic control, and (4) to geometrically determine the geoid height.

GNSS station criteria are different from that of conventional surveys—no inter-visibility necessary but sky obstructions should be avoided or minimized. A minimum of two receivers are required to survey and adjust baselines. If using a single GNSS receiver, data from a CORS network station can equate to a second GNSS receiver. The network is built up baseline by baseline, and each *baseline* or *session* is independent. A minimum of two receivers typically means a network can be observed by GNSS *radiations* or *leap-frog traversing*. And although structural and logistical factors will influence progress, speed of station coordination must be maximized, especially for detail and engineering surveys.

GNSS surveys for geodetic control must be performed to far more rigorous accuracy and quality assurance standards than those for control surveys for general engineering, construction, or small-scale mapping purposes.

Geodetic control stations are substantially monumented so that they will be both stable and durable. To support precise positioning, monuments must be stable and protected, minimizing movement due to frost, soil conditions, crustal motion, and human disturbance. They must be recoverable for future use.

6.4.2 STATION LOCATIONS, DISTRIBUTION, AND ACCESS

In traditional networks, stations are located on higher grounds due to optical inter-visibility requirements (Figure 6.4a). In a GNSS network, stations are located where needed (Figure 6.4b), and terrain does not influence site selection.

The accuracy of the survey network may vary considerably depending on the spatial distribution of the points used (Figure 6.5). For example, in an adjustment for transformation parameters, points should not be collinear because components of rotations about axes parallel to the line of points cannot be determined. For a stable solution, it is important that the network is well distributed spatially. For example, a network with uneven geographical spread of points will bias the solution toward the areas of high density. This often causes points in areas of low density to have large corrections to their coordinates.

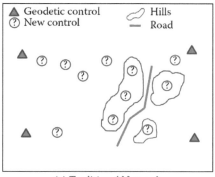

(a) Traditional Network

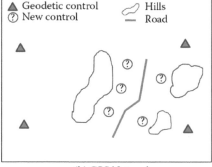

(b) GPS Network

FIGURE 6.4 Selecting GNSS station locations. Terrain need not influence site selection in a GNSS network. (a) Traditional Network, (b) GPS Network.

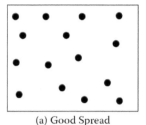
(a) Good Spread

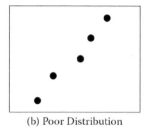
(b) Poor Distribution

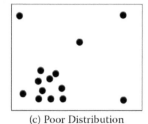
(c) Poor Distribution

FIGURE 6.5 Survey network geometry. (a) shows good network geometry, (b) and (c) show poor network geometry.

Having said this, spatial distribution of network points may be determined by two special case scenarios:

1. *Designing a new network:* In this case, there is full control in terms of where the points should be located on the ground. The criteria might be based not only on survey standards and guidelines but also on environmental and other related factors. For instance, a GNSS survey station requires clear sky visibility (i.e., areas with minimal GNSS signal obstructions). With careful site selections, a GNSS control network should aim to minimize use of areas with heavy urban canyon (tall buildings and skyscrapers) and thick foliage (tall trees, heavy forests, etc.). Alternatives may include, for example, use of building rooftops for station locations. When appropriate, the plan should also take into consideration any administrative or social requirements in accessing a project location. Some areas may require permission to access, while other areas may be physically impossible to access. Use of outdated information (e.g., outdated aerial photos) to plan project points may be unreliable—for example, a point may be mistakenly designed in the middle of a sewerage treatment plant or a private residential house.
2. *Resurvey of an existing network:* In the case where a network already exists, there may be very little that can be done to change the geometry and distribution of points. If, for instance, the network distribution is such that some parts have higher density than others, a more prudent approach may have to be applied in the resurvey. Areas with high density may be treated as subnetworks of the project so that in a single large project we may end up with two or more subnetworks, each with good geometry.

6.4.3 PLAN OF PROJECT POINTS

GNSS station locations are not dependent on optics that require station intervisibility, but rather on the ease of access. A minimum of three horizontal control stations for a given project area is recommended by the FGCC. Many more are usually required in a GNSS route survey. In general, the more well-chosen horizontal control stations that are available, the better.

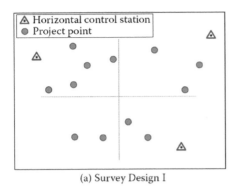
(a) Survey Design I

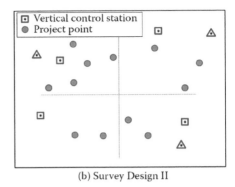
(b) Survey Design II

FIGURE 6.6 Survey design. (a) Project with horizontal control. (b) Project with both horizontal and vertical control.

The location of control stations relative to the project area is an important consideration (Figure 6.6). For work other than route surveys, the rule of the thumb is to divide the project area into four quadrants, and to choose at least one horizontal control in each quadrant. The absolute minimum for the actual survey is to have at least one horizontal control in three of the four quadrants. The control points should be as near as possible to the project boundary, and supplementary control can be added in the interior for more stability.

For route surveys, the minimum should be one horizontal control at the beginning, the end, and the middle of the route. Long routes should be supplemented with control on both sides of the line at appropriate intervals.

A minimum of four vertical controls is recommended, although a large project should have more. The vertical controls are better located at the four corners of the project area. Route surveys require vertical control at the beginning and the end, supplemented at intervals for long routes.

It should be clarified here that the concept of control referred to in Figure 6.6 does not necessarily mean that the control stations already have known (fixed) coordinates prior to the survey. In most cases, the project is such that each of the control points is surveyed for the first time together with the rest of the project points, and then held fixed during a network adjustment. Alternatively, the data from control points could be preprocessed separately with data from CORS and the resulting coordinates taken to be the fixed values.

6.4.4   DESIGN BY LEAST SQUARES AND SIMULATION

In traditional surveying, angles and distances are observed. In GNSS positioning, ranges to satellites are observed. In both methods, the final parameters are point coordinates. The errors in observations (angles, distances, and ranges) are propagated into the final coordinates. A presurvey analysis is sometimes necessary to determine what quality of observations is needed (e.g., $\pm 1''$ or ± 1 cm) or how many times each measurement should be repeated. These decisions can be made so that the accuracy of your results meets the level you want or satisfies the project specifications.

This can be done before the fieldwork unless the project is very small, similar to a previous one, or you are under instruction to do certain observations in a certain way.

In a design by least squares, the procedure is

1. Estimate or guess the variances of observations before making them, and then
2. Calculate the quality of the parameters (in this case the coordinates).

For any type of survey network design, reasonable values for the standard deviations of the observations can be based on instruments, techniques, or number of repetitions. In the case of GNSS, there will be errors in GNSS observed ranges that will propagate into the position coordinates. These will then affect the network adjustment involving "observed" GNSS baselines (Section 6.5.1).

The error budget in a GNSS observed range consists of satellite orbit errors, atmospheric errors, multipath, the satellite clock, and receiver clock errors. With the assumption that these systematic errors can be removed, the remaining errors that can affect the position quality are the instrument's random noise (precision) and the site-dependent satellite geometry, or GDOP. This will be the fundamental assumption in this section of the book.

In order to estimate the quality of the parameters, a least squares computer estimation program is applied. It is usually possible to use the same computer program that will do adjustment of the real observations. The input includes approximate values of the observations and estimates of their expected quality (i.e., standard deviations). In such a program, the following can be calculated without real observations:

Estimated quality of parameters:	$Q_x = [A^T P A]^{-1}$
Estimated quality of adjusted observations:	$Q_L = A Q_x A$
Estimated quality of residuals:	$Q_v = Q - Q_L$

The A matrix is calculated from approximate coordinates, and P is the inverse of Q, which is based on the input standard deviations.

A:	design matrix of partial derivatives w.r.t. parameters
P:	inverse of input covariance matrix (VCV) of observations
Q:	VCV matrix of observations

Optimization by simulation works on the assumption that there are no systematic errors in the observations, and also that observations are as good as the input standard deviations state.

A best guess for quality of GNSS position coordinates for a given project point can be made from the precision of the instrument that will be used to observe that particular point. The precision is normally stated according to the type of GNSS method to be applied (e.g., static, rapid-static, or kinematic). Such information is usually available from the instrument's manufacturer, although they can also be obtained from other sources such as the *GPS World Magazine* that usually conducts *Annual GPS Receiver Surveys*. For a GNSS network simulation, the parametric model equation will be a baseline equation in which the baseline vectors will be the observations, and the coordinates of all points in the network will be the parameters to be adjusted (estimated).

Here is the suggested procedure:

1. First, the standard deviations of each point (as estimated from the type of GNSS instrument and method) can be scaled on the basis of individual site characteristics. In other words, in a given project area different points will have different site obstructions, hence different GDOP characteristics. Exactly how a GDOP value can impact on the quality of site coordinates is formulated in the next section.
2. Second, assuming the baseline vectors to be the observations for a network adjustment, the quality (standard deviations) of a particular baseline vector will be an error propagation from the standard deviations of the two points defining that particular baseline.

The GDOP consideration in the first step might not be necessary if the project is located in an entirely open sky area.

6.4.4.1 Computation of GDOP and Its Effect on the Position Results

If (X_r, Y_r, Z_r) are the ECEF coordinates for a receiver r, and (X^s, Y^s, Z^s) are the ECEF coordinates for a satellite s, the range ρ_s between the receiver r and the satellite s is given by:

$$\rho_s = \sqrt{\left(X^s - X_r\right)^2 + \left(Y^s - Y_r\right)^2 + \left(Z^s - Z_r\right)^2}.$$

Directional partial derivatives w.r.t. the receiver's X, Y, Z and time (t) are

$$\frac{\partial \rho_s}{\partial x} = -\frac{\left(X^s - X_r\right)}{\rho_s}, \frac{\partial \rho_s}{\partial y} = -\frac{\left(Y^s - Y_r\right)}{\rho_s}, \frac{\partial \rho_s}{\partial z} = -\frac{\left(Z^s - Z_r\right)}{\rho_s}, \text{ and } \frac{\partial \rho_s}{\partial t} = -1.$$

The minimum number of satellites required to compute the receiver position is four. Therefore, assuming four satellites are observed by r, the strength of the geometry formed by the four satellites can be estimated by the GDOP computation. The procedure is to: (1) compute the design matrix A from the partial derivatives for each of the satellites, and then (2) compute the covariance matrix from which the GDOP can be estimated. Thus,

$$A = \begin{bmatrix} a_{x1} & a_{y1} & a_{z1} & a_{t1} \\ a_{x2} & a_{y2} & a_{z2} & a_{t2} \\ a_{x3} & a_{y3} & a_{z3} & a_{t3} \\ a_{x4} & a_{y4} & a_{z4} & a_{t4} \end{bmatrix}, Q = \left(A^T A\right)^{-1},$$

$$GDOP = \sqrt{Q_{11} + Q_{22} + Q_{33} + Q_{44}},$$

where

$$a_{xs} = \frac{\partial \rho_s}{\partial x}, \ a_{ys} = \frac{\partial \rho_s}{\partial y}, \ a_{zs} = \frac{\partial \rho_s}{\partial z}, \text{ and } a_{ts} = \frac{\partial \rho_s}{\partial t}.$$

The uncertainty of the GNSS receiver position is scaled (increased) by the GDOP factor. The approximation of the effect is given as follows:

$$\sigma = GDOP\sigma_0,$$

where
  σ = the uncertainty of the position (parameters)
  $GDOP$ = the geometric dilution of precision factor
  σ_o = the uncertainty of the measurements (observations)

6.4.4.2   Use of Predicted GDOP Maps

An up-to-date GDOP map predicted using an accurate 3D GIS model could offer a cost-effective alternative to individual site visits for checking satellite obstructions, especially for a large network with numerous points. Such GDOP maps would facilitate GNSS survey network design by simulation, in which the predicted GDOPs at point locations are applied to scale standard deviations of inputs. A software designed to predict quality of station positions on the basis of GDOPs mapped using some kind of a 3D GIS model would be ideal. Such a 3D GIS model could come from LIDAR data, satellite imagery, or aerial photos. It could even be based on use of Google Earth utility.

Basically what is needed to predict GDOPs in a project area is to generate obstruction surfaces. To generate obstruction images, all that is needed is an obstruction surface and a terrain surface. These are most easily extracted from multiple-return LIDAR data but could, in theory, be extracted from any elevation model, provided the heights of objects in the model are sufficiently accurate. Currently LIDAR data are only available for a small percentage of areas but the approach could potentially be investigated to accept data from other sources, and hence becoming much more useful to surveyors.

6.4.4.3   Network Design Terminology

The following terminologies are commonly used in a survey network design:

1. *Zero-order design (datum problem):* Which points or lines to hold fixed
2. *First-order design (network configuration problem):* The geometric layout of observations
3. *Second-order design (weight problem):* The quality of observations and how many repetitions
4. *Third-order design (densification problem):* How to incorporate your observations into existing survey control.

To work on the network configuration problem, change the design matrix *A* of partial derivatives. The partial derivatives are those with respect to each of the three GNSS baseline vector components in the *X*-, *Y*-, and *Z*-directions.

The second-order design (weight problem) is implemented through the input of standard deviations for the observations, as explained earlier. In other words, it is simply the design or change in the design of the weight matrix *P* (inverse of the VCV of observations, *Q*).

6.4.4.4 Summary of Procedures for Design by Simulation

The following are the procedures for survey network design by simulation as outlined in Harvey (2006):

- Use your knowledge of the topography, available instruments and techniques, and cost and logistics to design a first draft of the plan of possible measurements.
- Determine approximate coordinates of the points.
- Select reasonable values for the standard deviations of the observations based on instruments, techniques, and number of repetitions.
- Put all this information into a least squares program and run it in simulation mode.
- Interpret results.

Check that the standard deviations and error ellipses of coordinates are small enough to satisfy the project requirements. If satisfied with the results, consider the costs and logistics of field measurements. If the results are not good enough, then if possible, improve the quality of some observations. In any case, consider different scenarios as appropriate, for example, when to add or remove observations, when to increase or decrease the number of repetitions, whether to use a different instrument of better or lesser quality, and so forth.

6.5   GNSS OBSERVATION PLANNING AND OPTIMIZATION

6.5.1   Forming GNSS Baselines and Loops

A GNSS baseline is formed when two receivers observe the same set of satellites simultaneously. Therefore, in a network with many project points, several baselines can be formed sequentially by each time observing two project points simultaneously (Figure 6.7). *Independent baselines* are the minimum number of baselines necessary to observe the entire network to produce a unique solution. Redundant baselines, if observed, will form *loops* within the network.

Other terminologies used with respect to baseline formation include trivial, dependent, and nontrivial baselines. *Trivial* or *dependent* baselines are those that can be defined over and above the minimum number of necessary independent baselines for a unique network solution. For instance, in a three-station network, two baselines will be independent and the third will be trivial. *Independent* baselines are also known as *nontrivial* baselines.

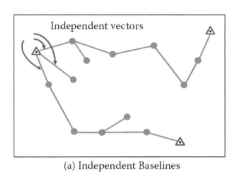

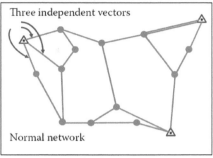

(a) Independent Baselines   (b) More Redundancies, Better Design

FIGURE 6.7 Forming baselines and loops. (a) Only independent baselines contribute to *network strength*, (b) Redundant baselines form loops.

Only independent baselines contribute to the geometric strength of the network. However, a better design incorporates use of redundant baselines observed to form *closed loops*. The concept of *loop closure* (or *loop misclose*) is applied to evaluate the internal consistency of a GNSS network. It can be described as follows. We first define a *GNSS session* as the duration in which a group of receivers (two or more) observe one or more baselines. A series of baseline vector components from more than one GNSS session, forming a loop or closed figure, are added together. The loop misclose is the ratio of the length of the line representing the combined errors of all vector's components to the length of the perimeter of the figure. Any loop closures that only use baselines derived from a single common GNSS session will yield an apparent error of zero, because they are derived from the same simultaneous observations.

In observing a network using only two receivers, each station would have to be occupied at least twice in order to connect every station with its closest neighbor. However, in the real world, most projects are usually done with more than two receivers. On the other hand, if it were possible to occupy all the project points with different receivers simultaneously, and do the entire project in one session, a loop misclose would be absolutely meaningless. Therefore, there are two aspects to the concept of redundancy in a GNSS network observation—repetition of independent baselines and reoccupation of stations. While it is not possible to repeat a baseline without reoccupying its endpoints, it is possible to reoccupy stations in a project without repeating a single baseline.

6.5.2 FINDING THE NUMBER OF GNSS SESSIONS

The minimum number of sessions n to observe a network of s stations using r receivers is given by (Hofmann-Wellenhof et al., 2001):

$$n = \frac{s - 0}{r - 0}, \text{ iff } o \geq \text{ and } r > o,$$

where o denotes the number of overlapping stations between the sessions. In the case of a real number, n must be rounded to the next higher integer.

Another approach for the design is where each network station is occupied m times. In this case, the minimum number of sessions will be:

$$n = \frac{ms}{r},$$

where n again must be rounded to the next higher integer.

The number s_r of redundant occupied stations with respect to the minimum overlapping $o = 1$ is given by:

$$s_r = nr - \left[s + (n - 1)\right].$$

These calculations do not include human error, equipment breakdown, and unforeseen difficulties. In other words, it is rather impractical to assume that a project would be trouble-free. The FGCS proposes the following formula for a more realistic estimate of the number of sessions:

$$n = \frac{sq}{r} + \frac{(sq)(p-1)}{r} + ks,$$

where the additional variables q, p, and k have the following meanings:

The variable q is a representation of the level of redundancy in the network, based on the number of occupations on each station.

The p variable, also known as the *production factor*, symbolizes the experience of a firm doing the project. A typical production factor is 1.1.

The variable k is known as the *safety factor*. A safety factor of 0.1 is recommended for GNSS projects within 100 km of a company's office base (OB). Beyond that radius, an increase to 0.2 is advised.

The computation of q requires a little more explanation. For example, in a network of 14 stations, assume that a survey design includes more than two occupations in ten of the stations. Further, assume that the total number of occupations planned to observe this network is 40, and that 4 receivers will be used in a minimum of ten planned GNSS sessions. By dividing 40 occupations by 14 stations, it can be estimated that each station in the network will be visited an average number of 2.857 times. This would represent the level of redundancy in the project. Hence, $q = 2.857$ in the FGCS formula.

For illustrations presented in Figures 6.8–6.10, the simple formulas can be applied to compute the *minimum number of sessions* as follows:

For Figure 6.8(a): $n = \dfrac{s - o}{r - o} = \dfrac{6 - 1}{2 - 1} = 5$ sessions

For Figure 6.9: $n = \dfrac{ms}{r} = \dfrac{(1)(2) + (3)(2) + (2)(2)}{3} = 4$ sessions

For Figure 6.10: $n = \dfrac{ms}{r} = \dfrac{(1)(2) + (3)(2) + (2)(2)}{4} = 3$ sessions

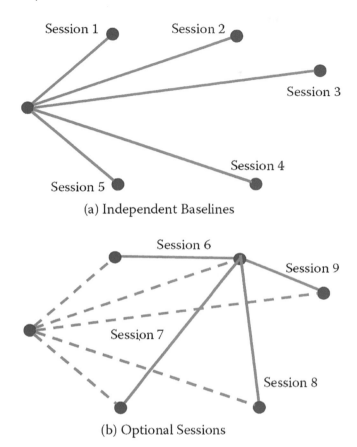

(a) Independent Baselines

(b) Optional Sessions

FIGURE 6.8   Observing sessions for 2 GNSS receivers. (a) Independent baselines, no trivial lines; (b) Optional sessions provide redundant checks.

6.5.3   GNSS Optimization

Having defined the number of sessions, the next problem in a network observation is to determine an optimized session schedule. A *session schedule* is defined as a sequence of sessions to be observed consecutively (i.e., the order in which sessions will be observed). If *n* represents the number of sessions for a given network, then the *number of possible session schedules* is given by n! This will be a very large number for some networks given that projects typically deal with networks comprising many points.

In operational research, complex combinatorial algorithms can be applied to solve for an optimized GNSS session schedule given a list of sessions to be observed and the cost to move receivers between points in the network. Details of such algorithms can be found in other publications, for example in Saleh (2002). Here, the idea of *GNSS network optimization* will be explained with the aid of a simple four-point network (Figure 6.11), as is also found in Dare and Saleh (2000). The goal of a GNSS network optimization is to search and determine the most suitable solution for

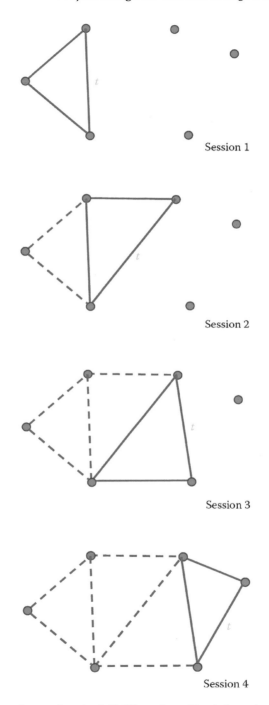

FIGURE 6.9 Observing sessions for 3 GNSS receivers. Two independent baselines per session, one trivial baseline (*t*) per session.

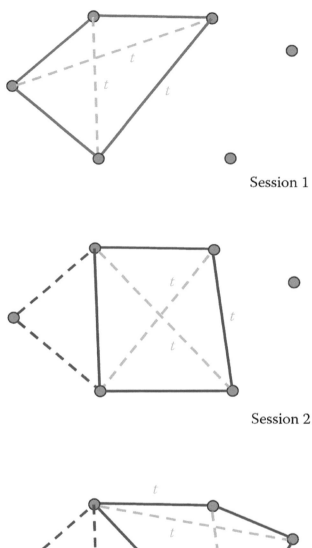

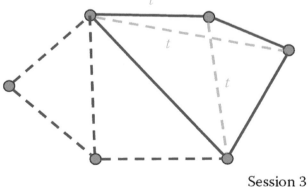

FIGURE 6.10 Observing sessions for 4 GNSS receivers. Six independent baselines per session, three trivial baseline (*t*) per session.

optimizing (minimizing or maximizing) an objective function (cost, accuracy, time, distance, etc.) over a discrete set of feasible solutions.

The four-point network in Figure 6.11 shows all the possible baselines (sessions) that can be measured (six in total) without repeating any observations. For two receivers, consider two schedules I and II as shown in Table 6.1, that have been arbitrarily selected out of a possible 720 (i.e., $n! = 6!$). Schedule II is apparently less efficient due to both receivers being moved between sessions.

If the numbers in Figure 6.11 represent the cost of moving a receiver in either direction between the two points, then the costs of the two arbitrary schedules are as shown in Table 6.2. For example, the cost of moving a receiver both ways between A and C is 6 units (i.e., from point A to C is 6 units and likewise for moving from C

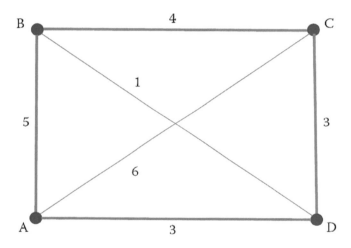

FIGURE 6.11 Simple four-point network producing a symmetric cost matrix. (Adapted from *Journal of Geodesy* 2000, 74: 467–478, Springer, Sept. 1, 2000, Figure 1, 468. Used by permission.)

TABLE 6.1
Session Schedules for Two Receivers

	Schedule I			Schedule II	
Session	Receiver 1	Receiver 2	Session	Receiver 1	Receiver 2
1	A	B	1	A	B
2	A	C	2	C	D
3	A	D	3	A	C
4	B	D	4	B	D
5	C	D	5	D	A
6	C	B	6	C	B

Note: Schedule II is less efficient.

TABLE 6.2
Comparison of Schedule Costs

	Schedule I			Schedule II	
Session	Receiver 1	Receiver 2	Session	Receiver 1	Receiver 2
1	—	—	1	—	—
2	0	4	2	6	1
3	0	3	3	6	3
4	5	0	4	5	3
5	4	0	5	1	3
6	0	1	6	3	5
Total	9	8	Total	21	15

TABLE 6.3
Symmetric Cost Matrix Generated Using Data in Figure 6.11

	A	B	C	D
A	0	5	6	3
B	5	0	4	1
C	6	4	0	3
D	3	1	3	0

to A). Whether the costs represent time, distance, or money is not important at this stage, as the design process is being described in a general sense. It is clear from these two schedules that the first option is the cheaper of the two. The challenge in the optimization problem is to determine, out of the possible 720 session schedules, the one particular schedule that will give the lowest cost from a specific cost matrix.

The receiver movement costs between all the neighboring points in the network can be represented by a cost matrix, where each element in the matrix is a cost between two points. If the cost of moving between two points is independent of the direction of the move (e.g., same cost whether moving from A to C or C to A), then a symmetric cost matrix is defined. The network in Figure 6.11 will produce a symmetric matrix as shown in Table 6.3, but this can be modified to produce a nonsymmetric matrix.

In Figure 6.12, the arrowed arcs represent the direction of movement along a line while the non-arrowed lines have costs that are independent of the direction of the move. Thus, the cost of moving from point A to B is 5 units, while the cost of moving from B to A is 4 units. In practice, a nonsymmetric cost matrix is more realistic as movement between points usually require a combination of driving, walking, and uphill journeys are slower compared to downhill journeys. If a helicopter is used to move between points, then the symmetric cost matrix may be more appropriate. Information gathering and research such as through reconnaissance or interpretation from satellite imagery can be used to collect data to enable costs between project points to be calculated.

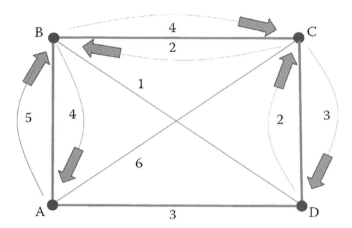

FIGURE 6.12 Simple four-point network producing a nonsymmetric cost matrix. (Adapted from *Journal of Geodesy*, 2000, 74: 467–478, Springer, Sept. 1, 2000, Figure 2, 469. Used by permission.)

The examples illustrated so far only indicate costs for moving between points. In practice, field operators will be moving from an office or hotel location to the project sites. Here we will refer to such a location simply as OB, and add an additional point to the problem description, as shown in Figure 6.13. The costs radiating from the OB represent the cost of moving a receiver from OB to the relevant point in the network.

There are then two case scenarios that can be considered in this problem. The first is that all the points in the network can be observed in a single working day. The second is that the project will take longer than one working day to complete, requiring multiple returns to the OB.

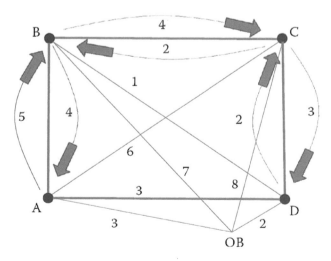

FIGURE 6.13 Expanded simple four-point network to include office base. (Adapted from *Journal of Geodesy*, 2000, 74: 467–478, Springer, Sept. 1, 2000, Figure 3, 469. Used by permission.)

In the first case scenario, an optimal route through all the points can be solved in an optimal manner using, for example, the traveling salesman problem (TSP), one of the many methods in operational research. The TSP can be stated as follows: Find a route that a salesman has to follow to visit n cities once and only once while minimizing the distance traveled. The GNSS network problem fits into this definition where the cities represent points.

In the second case scenario, the TSP is invalidated because it only allows each city to be visited once. To accommodate the need for multiple returns to the OB, a more complex multiple TSP (MTSP) is appropriate. In the MTSP, there is more than one salesman and they have to share visits to the cities in an optimal manner. The network shown in Figure 6.14 can be expanded to allow for this situation. The original OB is now represented by OB_0 and an additional return to the OB is represented by OB_1.

The concept of *nodes* and *arcs* is commonly used to allow use of algorithms that solve the MTSP for problems that are of more complex form. The points become known as nodes and the cost of moving between points in one or both directions become known as arcs. For example, in Figure 6.14, A, B, C, and D represent physical points on the ground. OB_0 represents the OB for start and end of a working period, and OB_1 represents the OB for the start and end of a different working period. Both OB_0 and OB_1 may be the same physical location, but they are represented as two separate nodes. Thus, the nodes now represent sessions and each arc the cost of moving receivers between sessions. A new cost matrix is constructed as shown in Table 6.4.

6.5.3.1 How Are the Elements of the Cost Matrix in Table 6.4 Interpreted?

Consider, for example, the element (AB, CD). This represents moving the receivers from session AB to CD. The receiver at point A has moved to C and the receiver at point B has moved to D. The cost of moving from point A to C is 6 units and the cost

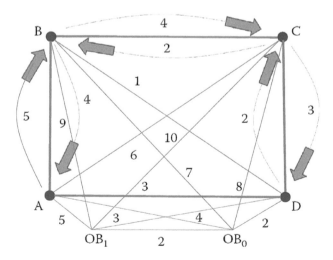

FIGURE 6.14 Expanded simple four-point network to include two working periods. (Adapted from *Journal of Geodesy*, 2000, 74: 467–478, Springer, Sept. 1, 2000, Figure 4, 470. Used by permission.)

TABLE 6.4

Initial Cost Matrix for Two-Receiver Problem with Two Working Periods

From/to	AB	BA	BC	CB	CD	DC	DA	AD	OB0	OB1
AB	∞	∞	5	6	6	4	4	1	7	9
BA	∞	∞	6	5	4	6	1	4	7	9
BC	4	6	∞	∞	4	1	6	4	8	10
CB	6	4	∞	∞	1	4	4	6	8	10
CD	6	3	2	1	∞	∞	3	6	8	10
DC	3	6	1	2	∞	∞	6	3	8	10
DA	5	1	6	5	3	6	∞	∞	3	5
AD	1	5	5	6	6	6	∞	∞	3	5
OB0	7	7	8	8	8	8	3	3	∞	2
OB1	9	9	10	10	10	10	5	5	2	∞

of moving from point B to D is 1 unit. The cost matrix is constructed to simulate the case of minimizing the time taken for the project. The rule used is that the cost to move between sessions is the maximum time of the individual movements. A different cost matrix could also have been constructed with the goal of minimizing, for example, the total distance (mileage) covered. In that case, the cost in element (AB, CD) would be the sum of the individual costs, giving 7 units. Whichever of the approaches is adopted, each session appears twice in the cost matrix to allow for all possible receiver movements. Thus, for session move AB to BC, there is also an allowance for possible move of AB to CB (i.e., the second receiver remains at point B, while the first receiver moves from point A to C). The block diagonal elements are set to ∞ (infinite cost) to prevent simple receiver swaps.

6.5.3.2 Calculating the Optimal Session Schedule

Using the cost matrix in Table 6.4, an optimal or near-optimal receiver session schedule can be calculated and plotted using a computer program implementing the operational research algorithms such as the MTSP. A number of such programs are currently available through the Internet and can be downloaded for free. An example of where to find such programs is the MATLAB® CENTRAL file exchange repository (see, e.g., files authored by J. Kirk). Numerous other examples exist that implement the MTSP algorithm in computer languages other than MATLAB. These include, for example, C/C++ and Fortran.

BIBLIOGRAPHY

Amolins, K. 2008. Mapping of GDOP estimates through the use of LIDAR data. Technical Report No. 259, University of New Brunswick.

Crossfield, J. 1994. GPS optimization. Class Notes.

Dare, P. and H. Saleh. 2000. GPS network design: Logistics solution using optimal and near-optimal methods. *Journal of Geodesy*, 74: 467–478.

FGCC. 1984. *Standards and Specifications for Geodetic Control Networks*. Rockville, MD: Federal Geodetic Control Committee.

FGCS. 1989. *Geometric Geodetic Accuracy Standards and Specifications for Using GPS Relative Positioning Techniques*. Rockville, MD: Federal Geodetic Control Subcommittee.

Harvey, B. R. 2006. *Practical Least Squares and Statistics for Surveyors*. Sydney, NSW, Australia: University of New South Wales. Monograph 13.

Hofmann-Wellenhof, H., H. Lichtenegger and J. Collins. 2001. *GPS: Theory and Practice*. 5th ed. New York: Springer-Verlag.

Lohani, B. and R. Kumar. 2008. A model for predicting GPS-GDOP and its probability using LiDAR data and ultra-rapid product. *Journal of Applied Geodesy*, 2(4): 213–222.

NGS Control Stations (NGSCS), http://www.metzgerwillard.us/ngscs.html (accessed November 22, 2009).

Ogaja, C. 2011. *Applied GPS for Engineers and Project Managers*. ASCE Press. ISBN 978-0-7844-1150-6, https://doi.org/10.1061/9780784411506.

Ogaja, C. 2022. GNSS CORS networks and data. In: *Introduction to GNSS Geodesy*. Cham: Springer. https://doi.org/10.1007/978-3-030-91821-7_6.

Rizos, C. 1997. *Principles and Practice of GPS Surveying*. Sydney, NSW, Australia: University of New South Wales. Monograph 17.

Saleh, H. A. 2002. Ants can successfully design GPS surveying networks. *GPS World*, 13(9): 48–60.

SPCS83 for Google Earth, http://www.metzgerwillard.us/spcge/spcge.html (accessed November 22, 2009).

USGS Quadrangles (QUADS), http://www.metzgerwillard.us/quads (accessed November 22, 2009).

Van Sickle, J. 2007. *GPS for Land Surveyors*. 3rd ed. Boca Raton, FL: CRC Press.

EXERCISES

1. Case Study: This project question requires that you identify (find) 30 existing survey monuments in a local area (e.g., within a 5-mile-by-5-mile area). You will then carry out the following tasks:

   a. Plot all the monument locations on an existing map or aerial photo or satellite image of the area. This should only give a rough indication of the point locations on such a map or image.

   b. Obtain existing coordinates of all the monuments and transform them into the GNSS datum.

   c. Visit all the sites and prepare a detailed obstruction diagram (also known as site visibility plot) for each site.

   d. Take four good quality digital photos for any two sites of the network, that is, per site obtain as follows (and capture part of the sky view, including any obstructions):
      - One photo with the view looking north
      - One photo with the view looking west
      - One photo with the view looking south
      - One photo with the view looking east

   e. Prepare a plot of the GDOP values for each site using a GNSS survey mission planning software.

   f. Estimate a single GDOP value for each site by averaging over a 12-hour window. If that value is infinitely big for a particular site, just assign a GDOP value of 20.

   g. Obtain the positioning accuracy (or precision) specification of a modern survey-grade GNSS receiver that you could have used to survey the entire network. This information can be obtained from the manual, vendor, or the *GPS World* magazine annual receiver surveys.

   h. Use the GDOP values obtained in (f) and the receiver precision obtained in (g) to estimate the scaled errors for each site of the network.

   i. By using a least squares computer program, perform a least squares simulation of the network design:

      – Use coordinates in the GNSS datum and scaled standard deviations of the sites as input to the program.

      – Prepare an adjusted network plot showing the applied corrections per site (e.g., in the form of error ellipses).

      – Overlay the adjusted network plot, with the error ellipses, on an imagery of the project area and give comments. Alternatively, overlay the network plot on an aerial view of a 3D GIS model of the area and give comments.

2. Case Study: In this question, you are required to provide surveying services to assist with the design of the wetland model in Figure 1.4. The primary goal of the project is to establish the coordinates of the well at point B and the four corners of the rectangular model. Each team has been assigned a set of starting coordinates for the point at well A (Table 6.5). The following tasks are required of each team:

   a. Plan and design of a GNSS control survey. You may include use of conventional methods where appropriate, with a justification.

   b. Demonstrate use of appropriate photogrammetric information.

   c. Demonstrate use of appropriate GIS tools and technologies.

   d. Demonstrate use of appropriate topographic information.

TABLE 6.5

Assigned SPC Coordinates of Well at Point A

Team	N (ft)	E (ft)	Zone
1	2196166.270	6398676.762	0404 CA4
2	2188910.138	6322611.883	0404 CA4
3	2207493.852	6343396.460	0404 CA4
4	1988901.680	6307117.867	0406 CA6
5	2184954.718	6294372.182	0402 CA2
6	2002948.691	6480702.211	0402 CA2
7	2462911.772	6424865.843	0401 CA1
8	2285616.844	6797601.040	0402 CA2
9	2267091.972	6413937.379	0403 CA3
10	1844002.690	6383762.930	0403 CA3

The work to be done consists of
- Research and information gathering; and
- Project planning and design (logistics, network plans, and observation procedures).

GNSS receivers to be used for the survey shall be survey-grade for cm-accuracy (*not* sub-meter receivers for mapping purposes). If conventional methods and equipment are to be used, the horizontal control closure shall meet or exceed 1:20,000.

Each project team should provide the following:

a. Planimetric and photographic information of project sites.
b. Assessment of physical accessibility of each project point.
c. Assessment of multipath conditions at each project point.
d. Assessment of satellite obstruction at each project point using station visibility diagrams.
e. Plot of approximate location of control stations within a 50-mile radius of one of the project points. For projects located within the United States, a NGS radial search may be done using the Google Earth. Indicate which of the points are from the GNSS continuously operating networks (CORS).
f. A network diagram clearly showing the site naming/numbering convention used.
g. Plan of observation schedules for the network diagram (indicative of baselines, sessions, and loops).
h. An optimized observation plan for each of the following scenarios:
   - One field personnel with a single GNSS receiver
   - Two field personnel, each with a GNSS receiver
   - Three field personnel, each with a GNSS receiver

   An optimized observation plan is the one that costs the minimum to observe the entire network. This should be clearly demonstrated with examples.

3. Your company intends to submit a proposal to establish a high-precision GNSS reference station network. This network will consist of five permanent GNSS tracking stations (stations A, B, C, D, and E) and cover an area of 15×15 miles. Your company has four GNSS dual-frequency receivers at its disposal.

a. How many sessions will be required to create the network? Include an explanation of the receiver deployment as part of your answer.
b. What will be the duration of each session?
c. How will you tie your network to an existing datum such as NAD83? Include the number of datum stations that will be included in this project as part of your answer.
d. In addition to your GNSS observations, name one type of information or hardware that can improve the precision of your GNSS baseline result.
e. You are required to include some form of terrestrial information, such as angles and distances, in your network. Explain how you will weigh the GNSS baseline results versus terrestrial data when doing the network adjustment on the project.

7 Topographic Surveys

7.1 GENERAL CONSIDERATIONS

7.1.1 Project Scope and Requirements

Topographic survey requests are often general in nature and often accompanied with a request for a cost estimate to perform the survey. In many cases, the survey details, site conditions, scope, and accuracy requirements are not specified; or, more often than not, the actual work required far exceeds the given budgeted amount. The burden is often placed on the surveyor to design a survey accuracy and density that will best satisfy the design requirements that the requesting entity desires. It is rare that the requesting user ever obtains the detail required for the project. Likewise, it is equally rare that the surveyor is able to perform the quality of survey he feels is necessary to adequately define the project conditions. In many cases, an advance site visit may be necessary to assess the actual conditions and provide a reliable budget estimate (time and cost) to the requesting client (Figure 7.1).

A control monument: Brass cap on concrete

FIGURE 7.1 An advance site visit would be essential in planning.

DOI: 10.1201/9781003297147-10

7.1.1.1　Sample Topographic Survey Request

The following excerpts describe a sample scope of work. The client's request may not have been as detailed as this version—it may have only requested a topographic site plan survey without any detailed map scale, accuracy, or utility requirements.
General Surveying and Mapping Requirements

1. General site plan feature and topographic detail mapping compiled at a target scale of $1'' = 50\,\text{ft}$ and 1 ft contour interval for the area annotated on exhibit. Collect all existing pertinent features; location of trees; fences, retaining walls, and driveways; buildings; and other structures; fire hydrants, drainage, and sewer; and any other visible features on the site. Collect all surface utility data and conduct a thorough search for evidence of subsurface utilities. An underground gas line runs through a portion of the site. Gas line markers are visible.
2. Set control monumentation to adequately control construction layout. Monumentation shall be set in an area outside the construction limits so as not to be disturbed during construction. Existing control monumentation within the vicinity may be used in lieu of setting new monumentation. All control monumentation, set or found, shall be adequately described and referenced in a standard field book.

This scope effectively describes the requirements of the survey. It does not specify all survey details that could be listed. For instance, it does not state what topographic elevation density is required on ground shot points. These types of details are usually left to the field surveyor to develop—presuming he knows the purpose of the site plan mapping project and is familiar with subsequent application requirements. It is therefore critical that the field survey crew be knowledgeable of the ultimate purpose of a project so that they can locate critical features that may impact the intended application.
An alternate method of describing topographic survey requirements is a checklist form.

7.1.2　OTHER GENERAL CONSIDERATIONS

7.1.2.1　Topographic Survey Planning Checklist

Upon receipt of a request for a topographic survey, as part of the planning process it is best to logically resolve many of the variables associated with the project. The following partial checklist (which is not in any particular order of preference) may be used as a general outline for that process.

a. *End-use of map or data:* How will the data or map be used? Site planning? Construction plans and specs? Geographic Information System (GIS) application? Will you count each tree, species, and size? Plot boundary? Compute lot areas?
b. *Project planning:* Thoroughly read request from client (request may be different than verbal agreement). What is the purpose of the survey? What did the site look like a year ago? What will it look like in 1 year?

c. *Site considerations:* Has the client or requestor walked the site recently? Have you personally walked the site recently? Safety hazards to consider are as follows:
   - Steep slopes
   - Busy roads
   - Road signs
   - High speed railroad
   - Poisonous trees, species
   - Weather patterns
   - Local mentality.
d. *Horizontal coordinate system:* UTM or local (e.g., state plane, on what zone?); true projection or ground surface? A perpetual coordinate system is usually better.
e. *Vertical coordinate system:* MSL or local? Perpetual vertical system (and conversion to datum) recommended; specify adjustment date if necessary.
f. *Stationing:* Stationing is a disjointed coordinate system where one axis is the STATION and the other is OFFSET, and the STATION axis rotates at every PI. Distances are usually ground distances, great for linear surveys such as roads, railroads, and levees.
g. *Units of measure:* Foot (U.S. or International Survey foot?), meter, mile, nautical mile, ground distances, or grid distances?
h. *Existing control:* Decide what horizontal and/or BM to hold—two or more. Have these monuments been reset? Always check between two or more existing monuments. Will you protect monuments during the construction phase or relocate them?
i. *Control monuments:* Set or reference permanent control. The same control should be used for project boundary recovered or set, map for design, plans and specs, construction, as-builts, reference ties, and digging permit? Set in protected place and set witness marks.
j. *Control survey:* Different procedure compared to mapping (see Chapter 3). Qualify monument coordinates with a level of accuracy and archive them.
k. *Control diagram:* Build for mapping, design, construction plans and specs, and archive; should include original and newly set monuments, coordinate system (grid, ground–ground distance/grid distance, combination factor and grid factor, etc.), and reference ties.
l. *Deliverables:* Topographic map: cross-section plots. Digital Terrain model: ink on mylar. Color: black/white. Digital files: digital file specification/format. Font size: line weights. Global origin: sheet size. Title block format: metadata. Field book: computation files. Daily reports: surveyor's report (pertinent data, relevant comments). Digital media type: CD, DVD, portable disk.
m. *Deliverable format:* CADD/GIS environment? Does file or map have to match existing data? What software will be used to view data? Will a variety of output files be required? File type: .DWG, .SHP, ASCII, .DXF, .TIF, .PDF; file size: Will the files be too large for your computing resources? Will the files match into existing software or computer or are you planning new software and computer?

n. *Schedule:* Is time critical? Will work be contracted? How long to advertise, select, and negotiate? Fiscal year (dated money)?

o. *Equipment and resources:* Will the available equipment be sufficient for project requirements or are you planning new equipment, software, and/or training?

7.1.2.2   Rights-of-Entry

When entering property to conduct a survey, rights of property owners should be respected. Permission to enter property must be acquired whenever necessary. While on the property, survey crew must adhere to any rules, regulations, directives, and verbal guidance as set forth by property owners or public authorities.

Survey monuments should be placed in such a way as to not obstruct the operations of property owners or be offensive to their view. Monuments set as a result of the survey should be set below ground level to prevent damage by or to any equipment or vehicles, such as grass-cutting tractors.

7.1.2.3   Sources of Existing Data

When a survey request is received, the first effort should be to research the project area to ascertain what types of useful geospatial data exist of the same site. However, given the highly detailed scale of topographic surveys and the need for current conditions, it is highly improbable that an archived survey of sufficient detail can be found. Regardless, existing control will still be needed to reference the topographic mapping. A variety of databases can be accessed to obtain horizontal and vertical control from various agencies. One or more of the following sources of geospatial data may need to be researched before performing a topographic survey:

- As-built drawing files
- Aerial photos
- Topographic maps
- Imagery: satellite (such as Google Earth) or orthophoto
- Records related to real property surveys
- Geodetic control: national (such as National Geodetic Survey [NGS] control) and local (state, city, and regional agency control)
- Utility maps: electric, gas, storm, cable, telephone, fiber optic, etc.

7.2   PROJECT CONTROL FOR TOPOGRAPHIC SURVEYS

Topographic surveys of facilities, utilities, or terrain must be controlled to some reference framework, both in horizontal and vertical. The reference framework should consist of two or more permanently monumented control points and/or benchmarks located in the vicinity of the project. Those control points can then be used to perform supplemental topographic surveys of the project. This concept is illustrated in Figures 7.2 and 7.3—control is brought in from two existing points using static GNSS observations. Four points, one in each quadrant of the project area, are positioned to establish a local baseline control. From these points, subsequent topographic detail is surveyed using either a total station or real-time kinematic (RTK) methods. Although

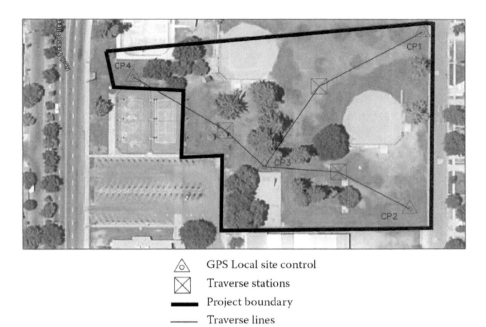

GPS Local site control
Traverse stations
Project boundary
Traverse lines

FIGURE 7.2   Proposed control at a project site.

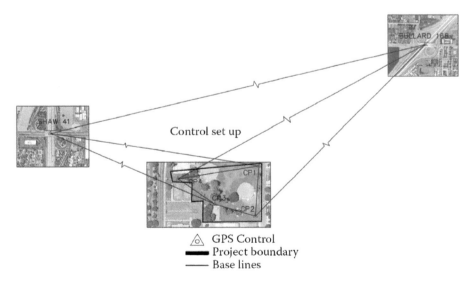

GPS Control
Project boundary
Base lines

FIGURE 7.3   Extending project control from existing points.

the control would be simply referenced to the satellite-based WGS-84 system, connections to local reference frameworks (such as using a UTM grid) can be made. Vertical control is usually established relative to the nearest existing benchmarks.

a. *Project control relative accuracy:* In general, horizontal and vertical accuracy of the control points used to control topographic surveys need only be to third-order, relative to themselves. In practice, if these control points in and around the project site have been interconnected by total station, differential leveling, or static/kinematic GNSS techniques, their relative accuracy will be far greater— upward of 1:50,000 to 1:100,000 closures. Positional accuracies within a project site should be around ±0.2 ft in X-Y, and better than ±0.1 ft in the vertical.

b. *Project control absolute accuracy:* The absolute accuracy of project control is the accuracy defined relative to some local, regional, or national reference framework. These frameworks might be the ones maintained by governmental agencies such as the NGS or international bodies such as the International GNSS Service. The National Spatial Reference System (NSRS) is an example framework that can be used for a project within the United States. Control by GNSS methods is increasingly common given the worldwide proliferation of continuously operating reference station (CORS) networks and related services (see Chapter 6). Maintaining a good relative accuracy with an adjoining project or control network is far more important than accurate connections to distant networks. Likewise, connections to adjoining property boundary monuments are significantly more critical than those to distant networks.

c. *Boundary control:* Topographic surveys involving real property boundaries must always be connected to established property corners or adjoining right-of-way boundaries. Locations of structures, buildings, roads, utilities, and so forth, are surveyed and mapped relative to the property boundaries. Likewise, stakeout of planned construction must be performed relative to the boundaries. Control framework coordinates may be placed on property corner marks, and subsequent stakeout work should not be performed relative to distant control.

d. *Local project control:* On some occasions, there is no existing horizontal or vertical control within the immediate vicinity of a project. Two options are available:

   – Perform detailed surveys relative to an arbitrary coordinate system established for the project. For example, set two permanent reference points and assume arbitrary coordinates of 5,000-10,000-100 (X-Y-Z) for one of the points.

   – Perform traverse, leveling, and/or a GNSS control survey to bring in control to the project site.

      The first option used to be more common but nowadays, with the ease of extending control with GNSS, it is fairly simple to establish some form of control with reference to WGS-84 ellipsoid.

7.2.1   ESTABLISHING CONTROL AT A PROJECT SITE

A variety of factors must be considered in deciding whether and how to connect project sites to an external coordinate network. These include the following:

- *Cost*: Bringing distant horizontal and vertical control to a project site can be costly, and may exceed the cost of performing the detailed topographic survey itself.

- *Policy*: Agency/organizational policy or regulations may mandate that all site plan surveys shall be referenced to a particular reference framework. If this is the case, then it is up to the surveyor to perform the connection in the most cost-effective manner.
- *Accuracy*: Horizontal and vertical accuracy of topographic features relative to the project control must be adequately defined. However, most project sites have no real requirement for rigorous connections to an external framework. For example, absolute externally referenced positional accuracies of a building would be adequate at ±10 ft in X-Y, and ±3 ft in the vertical, whereas its local accuracy relative to an adjoining property line would be around ±0.1 ft in X-Y, and floor elevations better than ±0.02 ft relative to local utility connections.
- *Distance from a control network*: The distance of control points or vertical benchmarks from the project site might have an impact on the cost. In particular, if a distant benchmark requires a lengthy level line to bring in accurate vertical control. On the contrary, more options are available for bringing in horizontal control to a project site, such as GNSS static options using CORS networks.

7.2.2 PROJECT CONTROL DENSIFICATION METHODS

Depending on many factors (some of which are listed in Section 7.2.1), the method and accuracy of bringing in project control can be designed. The following paragraphs identify some of the common techniques that can be employed in establishing horizontal and vertical control relative to an existing network.

a. *Horizontal control:* Horizontal control is most effectively connected to an external network by one of the following methods:
   - Traverse surveys
   - Static GNSS surveys
   - Kinematic GNSS surveys.

   Traverse surveys with a total station are practical if existing control is fairly close to the project site. If traverse surveys are impractical, then a static GNSS observation may be more appropriate. At least two points of the existing network should be occupied. Alternatively, a static GNSS survey could be conducted at a point set up on the project site (Figure 7.4) and using the CORS network to adjust the point. Because most topographic surveys require only third-order accuracy relative to an existing framework such as the NSRS, short-term (1- or 2-hour) GNSS observations are normally sufficient for extending control to a project site.

b. *Vertical control:* If vertical control is required to a higher accuracy than can be achieved using GNSS surveying techniques, then conventional leveling methods must be used. Depending on the distance of the level run, third-order methods are usually sufficient.

FIGURE 7.4   GNSS survey set up.

7.2.3   EXTENDING CONTROL FROM A LOCAL PROJECT OR NETWORK

Most topographic surveys are performed on sites where existing geodetic or bound-ary control is readily available. Depending on the distance of the control from the project site, either total station traverse or static GNSS surveys are used to establish local control. Vertical control will typically be brought in by running third-order lev-els from two existing benchmarks. If boundary surveys are required, then all prop-erty corners should be recovered and tied in as part of the survey.

7.2.4   EXTENDING CONTROL FROM A DISTANT NETWORK

Permanent networks of continuously operating GNSS receivers, CORS, can be used to establish control at virtually any place (see Chapter 6). The use of CORS eliminates the need to occupy full baselines with two receivers. A single GNSS receiver is set up at a primary control point in the project site, and 1- to 2-hour static GNSS observations are recorded. These observations become the end of any number of selected baselines using stations in the CORS network. Static GNSS observations made at a project site can be adjusted to any number of nearby CORS stations. In the continental United States, the NGS provides a user-friendly CORS Web site which is linked through ngs.noaa.gov.

Azimuth orientation at the topographic project site is easily performed as part of the process of bringing in CORS control. A second GNSS receiver is set up at a marked point, hundreds of feet away from the first GNSS point. GNSS observations over the short baseline are made concurrently with the CORS baseline connections. The fixed solution over this short baseline will provide adequate azimuth orientation for subsequent topographic work at the project site. (Note that a fixed solution is required over this baseline.) Either end of the baseline can be used to fix the CORS-derived X-Y-Z position. An absolute accuracy of 10–30 seconds over a 1,000 ft baseline would be adequate assuming the survey site is small and no real property connections are required. If the site has deeded boundary alignments (e.g., bearings shown along a road or boundary), then these deeded bearings may be used for azimuth reference. GNSS-derived azimuths would have to be corrected to fit the local orientation.

7.2.4.1   Using Online GNSS Processing Services

Free online GNSS services that make use of CORS networks are discussed in Chapter 3. They can be used to establish accurate horizontal and vertical control relative to a national datum such as the NSRS of the United States. Online Positioning User Service (OPUS), provided by the NGS, is an example of such services. It is accessed at the Web address: www.ngs.noaa.gov/OPUS. OPUS provides online baseline reduction and position adjustment relative to three nearby NGS CORS reference stations. It is simpler to operate because only the user's observed data need to be uploaded as opposed to downloading three or more CORS RINEX files. It can also be used as a quality control check on previously established control points. OPUS provides horizontal coordinate solutions in both ITRF and NAD83, and an orthometric elevation on NAVD88 using the current geoid model. An overall root-mean-square (95%) confidence for the solution is provided, along with maximum coordinate spreads between the three CORS stations for both the ITRF and NAD83 positions.

7.2.5   Approximate Control for an Isolated Project

When performing a topographic survey at a remote location (i.e., no existing control in the project's vicinity), the following options are available:

- Establish a local arbitrary coordinate system. For example, set and mark a primary point with X-Y-Z coordinates (e.g., 10,000-5,000-100 m or ft). It is recommended that the X-Y coordinates be sufficiently different to avoid potential confusion between them, and the coordinates should be such that negative coordinates will not occur over the project site.
- For azimuth orientation, set and mark a secondary point 500–1,000 ft away from the primary point.
- Establish the azimuth orientation between the two points (i.e., baseline) using either:
  - Arbitrary azimuth of 0.000 deg.
  - Estimated azimuth (scaled from map or photo)
  - Magnetic azimuth (from transit or handheld compass)
  - Astronomic observation (Solar or Polaris)

- − 8–15 minutes GNSS baseline observation, holding autonomous position at the primary end of the baseline, or
- − Gyroscope.
- • Perform the topographic survey relative to these two points. Assume a tangent plane grid for the distances; hence, no grid or sea-level corrections are applied to observed distances.

The next consideration is whether a *nongeoreferenced* or a *georeferenced* control will be implemented in the final product of the topographic survey. The decision will mostly depend on the purpose or needs of the project for which the survey is being conducted. Here are some case descriptions:

a. *Nongeoreferenced control:* Georeferenced control is rarely required for construction projects; an arbitrary control described earlier would be adequate for all design, stakeout, and construction. In such cases, an arbitrary grid system can be established in minutes. The baseline is quickly marked with stakes, hubs, rebar, or nails at each end. Topographic surveys using a total station or RTK method can then be conducted, starting at one end of the arbitrary baseline. If needed, supplemental control traverses can be run to set additional marked control points around the project site. Optionally, RTK radial control points can also be set relative to the baseline.

b. *Approximate georeferencing using autonomous GNSS:* If georeferenced control is required on the isolated project, then autonomous GNSS positioning could be used to establish approximately georeferenced coordinate at the primary control point. All data observed on the arbitrary grid system can then be later translated and rotated to a georeferenced coordinate system. If only approximate georeferenced control is required, then a handheld GNSS receiver is adequate, and in that case it should be noted on survey records that the resultant coordinates are approximate. A few minute visual recording of the position is sufficient. A quick autonomous GNSS position on the other end of the baseline can be used to establish a rough geodetic azimuth (± 1 deg at best) of the baseline. If the receiver can convert Lat/Long to the local UTM zone, then the UTM coordinate system may be used to reference the project.

c. *Precise georeferencing using long-term static GNSS:* If a more precise georeferencing is required, then long-term (1–2 hour or longer) static GNSS observations can be made at the primary control point. With geodetic quality receivers, a high accuracy (better than ± 0.5 m) WGS84 3-D position will be obtained. If two geodetic receivers are available, a fixed solution is possible over the short baseline with only a few minutes of static observations.

d. *Coordinate transformations:* All drawings should clearly note the approximate georeferencing of the project, the method by which it was performed, and the estimated accuracy of the primary reference point. Previous topographic observations on an arbitrary coordinate system may be transformed to the WGS84/UTM grid or any other appropriate local grid system, using standard transformation routines found in most software packages. These routines will also apply grid- and sea-level corrections during the transformation, assuming they are significant.

7.3   MAP SCALE AND CONTOUR INTERVAL

It is absolutely essential that surveying and mapping specifications originate from the functional requirements of the project, and that these requirements be realistic and economical. In other words, specifying map scales or accuracies in excess of those required for the end goal of the project (design, construction, GIS, mapping, etc.) results in increased costs and may delay project completion. If the project is being done by and for an organization, it may be possible that a general guidance for determining map scales, feature location tolerances, and contour intervals is already available within the organization. However, if the project is being done privately, the surveyor should confer with the requestor to stay within budget. Table 7.1 is example of a general guidance from which specifications may be developed.

a. *Target scale and contour interval specifications.* Map scale is the ratio of the distance measurement between two identifiable points on a map to the distance between the same physical points existing at ground scale. The selected target scale for a map or topographic survey plan should be based on

---

TABLE 7.1

Sample (Hypothetical) Guidelines for Accuracies and Tolerances for Feature and Topographic Detail Plans

Project/Activity	Target Map Scale	Feature Position Tolerance	Contour Interval	Survey Accuracy
General construction site plans	1:500 40 ft/in	100 mm 50[a] mm	250 mm 1 ft	3rd order
Surface/subsurface utility plans	1:500 40 ft/in	100 mm 50[a] mm	N/A	3rd order
Building or structure design drawings	1:500 40 ft/in	25 mm 50[a] mm	250 mm 1 ft	3rd order
Grading and excavation plans (roads, drainage, curb, gutter, etc.)	1:500 30–100 ft/in	250 mm 100[a] mm	500 mm 1–2 ft	3rd order
Recreational site plans (golf courses, fields, etc.)	1:1000 100 ft/in	500 mm 100[a] mm	500 mm 1–2 ft	3rd order
GIS (housing, schools, boundaries, etc.)	1:5000 400 ft/in	10,000 mm N/A[a]	N/A	4th order
Archeological site plans and details	1:10 10 ft/in	5 mm 5[a] mm	100 mm 0.1–1 ft	2nd-I/II order

[a] Vertical.

---

the detail necessary to portray the project site. Surveying and mapping costs will normally increase exponentially with larger mapping scales. Therefore, specifying a too large scale or too small contour interval than needed to adequately depict the site can significantly increase project costs. Topographic elevation density or related contour intervals should be specified consistent with the existing site gradients. Photogrammetric mapping flight altitude or ground topographic survey accuracy and density requirements are determined from the design map scale and contour interval provided in the contract specifications.

b. *Feature location tolerances:* Feature tolerances should be determined from the functional requirements of the project (construction, stakeout, alignment, GIS mapping, etc.). It establishes the primary surveying effort necessary to delineate physical features on the ground (Figure 7.5). In most instances, a feature may need to be located to an accuracy well in excess of its scaled accuracy on a map—hence feature location tolerances should not be used to determine the required scale of a drawing or determine photogrammetric mapping requirements. Positional tolerances (or precision), such as in Table 7.2, are defined relative to adjacent points within the confines of a specific project area, map sheet, or structure, not to the overall project boundaries. For example, two catch basins 200 ft apart might need to be located to 0.1 ft relative to each other, but need only be known to ±100 ft relative to another catch basin 6 miles away.

c. *Maintaining relative precision in a project area:* All features located in a project should have the same relative position. However, in practice, the relative precision of features located farthest from the project control points will tend to have larger errors. To ensure that the intended precision does not drop below the tolerance of the survey, secondary project control loops or nets may be constructed from the primary project control network, depending on the size of the project. There may also be instances where trade-offs between survey control and scale may be necessary—dependent upon project costs and usable limits of scale.

d. *Optimum target scale:* The requesting client or surveyor should always use the smallest scale that will provide the necessary detail for the project. Once the smallest practical scale has been selected (e.g., on the basis of recommendations such as those provided in Table 7.1), determine if any other future map uses are possible that might need a larger scale.

e. *Optimum contour interval:* The contour interval is chosen based on the map purpose, required vertical accuracy (if any was specified), the relief of the project area, and also somewhat from the target map scale. Steep slopes would require an increase in the contour interval to make the map more legible. Flat areas will tend to decrease the interval to a limit that does not interfere with the planimetric details of the topographic map.

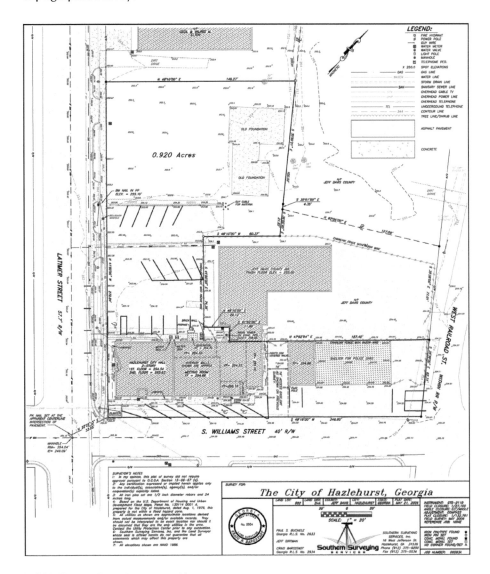

FIGURE 7.5 Sample topographic survey map.

BIBLIOGRAPHY

Crawford, W. G. 2002. *Construction Surveying and Layout: A Step-by-Step Field Engineering Methods Manual.* 3rd ed. West Lafayette, IN: Creative Construction Publishing, Inc.

Dewberry, S. O. 2008. *Land Development Handbook: Planning, Engineering, and Surveying.* New York: McGraw-Hill.

TABLE 7.2

Summary of National Map Accuracy Standards for Photogrammetric Mapping

Horizontal Scale	Feature Location (ft±)	Contour Interval (ft)	Vertical (90%) (ft)
1 in = 20 ft	0.5	1	0.5
1 in = 40 ft	1.0	2	1
1 in = 50 ft	1.25	2	1
1 in = 100 ft	2.5	2	1
1 in = 100 ft	2.5	5	2.5
1 in = 200 ft	5.0	2	1
1 in = 200 ft	5.0	5	2.5

Source: U.S. National Geospatial Data Standards.

EXERCISES

1. In a project case study of topographic surveying, discuss the general considerations for each of the following:
   a. Scope and requirements
   b. Extending project control
   c. Map scale and contour interval requirements
2. At what map scale would the difference between spherical and ellipsoidal coordinates be important (assuming that you can distinguish a line of 0.5 mm on a map)? Explain. Assume Earth's mean radius R = 6,371 km and ellipsoidal flattening of 1/300.
3. How many significant places would you need to specify latitude or longitude in decimal degrees for a 1 mm resolution map?
   a. 15
   b. 11
   c. 8
   d. 2
4. Which of the following is not an option for extending control in a topographic survey project?
   a. Continuously operating GNSS networks
   b. EDM traverse
   c. Precise leveling
   d. Airborne LIDAR
5. The output map scale is an important specification when designing a mapping project—it greatly affects the amount of detail and costs (or time). Which of the following statements is correct?
   a. The larger the scale map required (1″ = 100′ is smaller than 1″ = 200′) by a requesting client, the shorter it will take to produce and the less costly it will be.

    b. The larger the scale map required ($1'' = 100'$ is larger than $1'' = 200'$) by a requesting client, the shorter it will take to produce and the less costly it will be.

    c. The smaller the scale map required ($1'' = 100'$ is smaller than $1'' = 200'$) by a requesting client, the longer it will take to produce and the less costly it will be.

    d. The larger the scale map required ($1'' = 100'$ is larger than $1'' = 200'$) by a requesting client, the longer it will take to produce and the more costly it will be.

6. Mapping clients often fall short of meeting their goals by assigning the wrong product accuracy specifications to their project. Select the choice that correctly completes the following statement: Assigning a very strict product accuracy (such as Class I)

    a. Increases the amount of mapping data.

    b. Lowers the mapping scale.

    c. Decreases the number of contours.

    d. Implies larger contour interval.

8 GIS Application

8.1 INTRODUCTION

This chapter presents important considerations when planning or designing a GIS application. But first, what is a GIS application?

The following words exemplify what is a GIS application:

> With a simple database or spreadsheet, I could show you data on a famine in Africa but all you would see are the names of the countries and a bunch of numbers. With GIS I can make a map and show you where there is surplus food and where that surplus could be distributed (Unknown Author).

A *GIS application* is an automated process that generates a *spatially oriented* product or result needed by a user. GIS applications may include map update or map production, data query and display, spatial analysis, or other processes that use GIS software and geographic data. These applications give users in the office and in the field effective and easy-to-use ways to access information, answer questions, generate products, and support decision-making. For public sector organizations, GIS applications are integrated with other systems (e-government, document management, asset management, etc.) to support operations and serve the public.

GIS software packages come with a range of off-the-shelf functions that may be used without significant customization. Often, however, GIS users find it beneficial to use software configuration or development tools to create custom applications that support specific user needs. In this case, a *GIS (software) application development* work may involve any of the following types of customization: (1) configuration of new or modified graphic user interfaces, (2) automating access and integration with other systems, (3) developing "intelligent" forms to support efficient data entry, (4) developing application scripts for complex application workflows, (5) creating custom-design templates for map displays and reports, (6) creating a library of standard queries that can be accessed through a menu, and (7) programming analysis applications using GIS and other software development tools.

A GIS project may fall into one of the two categories listed or it may be a combination of elements from both a basic GIS application and a GIS software application development.

In a basic GIS application, the assumption is that a GIS system (software, hardware, people, communications network infrastructure, etc.) is already in place and software customization, if any, will be minimal. Therefore, the main project goal is to plan for research and preparation, data collection, analysis, and presentation. In some instances, a basic GIS application work could involve both the initial design and application of a GIS software system.

DOI: 10.1201/9781003297147-11

8.2   GENERAL CONSIDERATIONS

8.2.1   DATA REQUIREMENTS

Data requirements make up the most expensive part of implementing a GIS. Businesses, organizations, and government departments spend large sums of money on collecting, processing, and archiving geospatial data for GIS systems. Knowing the purpose of a project will ensure that research or fieldwork efforts are focused on collection of required features and attributes.

Illustration: Baseline inventory surveys capture overview data and provide basic information about many different features (Figure 8.1a). Specific application surveys capture specialized data and complex information describing a specific feature (Figure 8.1b). Most GNSS projects for GIS application are a combination of both baseline and specific application surveys. Familiarity with the client's (or organization's) use of data and how it supports their programs and business operations will help the decision process (e.g., in terms of choice of data collection and data conversion methods and tools, and use of appropriate data modeling and design tools).

The uses of a GIS product often dictate the type of data that will be needed and the accuracy of the data that must be obtained. From a management and cost-efficiency perspective, it would be important to identify all current and potential future projects that might use the data produced for the current project. This is called a needs assessment.

A needs assessment may range from a rather informal process of brainstorming with a few department heads to a very formal analysis conducted by a consulting firm; regardless of how the topic is approached, conducting a needs assessment is a very important phase in planning a GIS application project. Leaping directly to method(s) of choice, mapping scales, product accuracies, photo scales, pixel resolutions, and so on, puts the cart before the horse—the project result may be highly accurate and precise data may be used (but data that you discover down the road) that doesn't do exactly what you'd like it to do, or doesn't serve the needs of everyone in the project.

Illustrated in Figure 8.2 are some general questions that can be asked in a needs assessment for a GIS application development.

8.2.2   LEVEL OF ACCURACY

The purpose of the project determines the level of accuracy required. Accuracy is dependent on the GIS map scale. For example, a county map (Figure 8.3a) will imply higher feature location accuracy compared to a state map (Figure 8.3b), that is, large map scale versus smaller map scale will determine the method of choice for collecting the GIS feature location data. Although a GIS application will not have a fixed scale (data can be zoomed in or out), choosing a map scale is perhaps the most important component of the design for the following two reasons: (1) If a map and/or imagery is part of the GIS input data, the map scale will come with its inherent accuracy level and will also determine the limit of resolution for the resultant GIS product; and (2) the final GIS map scale will determine the method of choice for

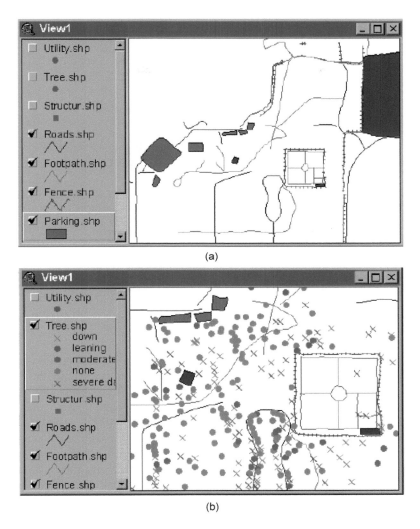

FIGURE 8.1 Deciding the purpose of the GIS. (a) Baseline Inventory, (b) Specific Application.

mapping feature locations. Table 8.1 gives an illustration of some typical GIS data sources and their respective levels of accuracy.

The output map scale is the first consideration and perhaps the most important specification when designing the project. The output scale determines the size of the output map for a defined geographic area and, most importantly, it determines the amount of detail that can be represented on or extrapolated from the map. Like accuracy, the output map scale greatly affects project costs and scheduling. The larger scale map that is required (i.e., $1'' = 100'$ is larger than $1'' = 200'$), the longer it will take to produce and the more costly it will be.

A PROJECT NEEDS ASSESSMENT
Who is involved with the project?   (Check all entities currently involved and those with potential involvement)   ☐ EMS ☐ Planning/Zoning ☐ Engineering ☐ Public Utilities ☐ Police   ☐ Fire ☐ School District ☐ Private Industry ☐ Other (please specify)
What are the primary intended applications?   (Check intended applications for the entities listed above)   ☐ Vehicle Dispatch & Response ☐ Mapping ☐ Terrain Modeling   ☐ Environmental Impact Assessment ☐ Engineering Design ☐ Crime/Accident   ☐ Traffic Planning ☐ Feature Identification ☐ Road Condition Mgmt.   ☐ Drainage Analysis ☐ Property Values ☐ Building Permits, Inspections, etc.   ☐ Facilities Management ☐ Other (please specify)
What are the primary data sets needed to achieve the goals?   (Check all that applies)   ☐ Digital Orthophotography ☐ DTM/Modeling ☐ Boundary Delineation   ☐ Utility Features ☐ Hydrographic Features ☐ Topographic Features   ☐ Structural Features ☐ Vegetation ☐ Transportation Features   ☐ Parcels ☐ Existing Data Conversion ☐ Other (please specify)
What software platform(s) will be utilized?   (Check all used by above agencies/departments)   ☐ ArcGIS ☐ ArcInfo ☐ ArcView ☐ Map Info ☐ AutoCAD ☐ Microstation   ☐ Other (please specify)

FIGURE 8.2   A project needs assessment.

The products that are needed as determined in a needs assessment (Figure 8.2) will dictate the map scale to be produced. Many features like street centerlines, edge of pavement, and buildings can be captured from many different map scales ranging from $1'' = 50'$ to $1'' = 400'$. Typical map scales for GIS mapping applications are $1'' = 100'$ for urban or developed areas and $1'' = 200'$ or $1'' = 400'$ for rural and less developed areas. However, should it be determined that features like fire hydrants or manholes are needed, an alternative such as $1'' = 50'$ or larger should be considered. It is important to select a mapping scale that ensures you can identify each feature you want to collect. Equally important is not to procure a scale that costs more without any practical benefit.

8.2.3   EXISTING INFORMATION

To support a specific GIS application, substantial amounts of new data will be gathered for its database. Most likely, however, some of the data will be obtained from existing sources such as maps, engineering plans, aerial photos, satellite images, and other documents and files that were developed for other purposes (Figure 8.4). These kinds of data could be available from past projects (for the client or organization) or externally from other sources. An important consideration in this regard, though, is whether the existing information is in paper or digital format. Digital base data may already exist for the project area (e.g., Figure 8.5) and appropriate data conversion methods should be planned for in cases where the required information only exists in the paper format.

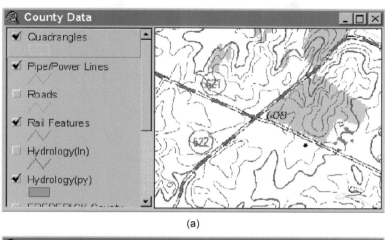

(a)

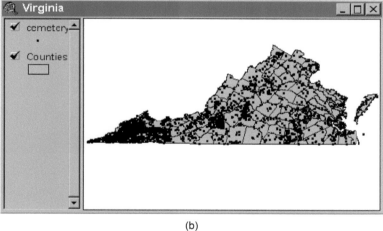

(b)

FIGURE 8.3 Level of accuracy. (a) County Data, (b) State Data.

TABLE 8.1
Levels of Accuracy

GIS Data Source	Accuracy (m)
1:250,000 USGS 1×2° series	±250
1:100,000 USGS 30×60 minutes series	±90
1:24,000 USGS 7.5 minutes quad maps	±12
1:2,000 site plans and tax parcel maps	±5
Mapping-grade GNSS	±1
Surveying-grade GNSS	±0.1
Laser transit boundary line survey	±0.05

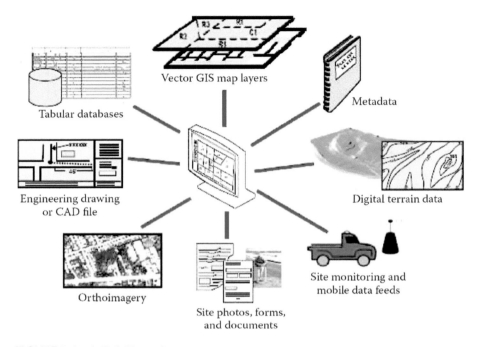

FIGURE 8.4   A GIS illustration.

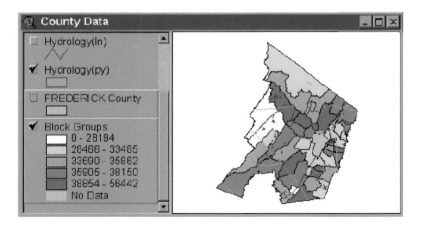

FIGURE 8.5   Existing data.

GIS information is increasingly available in the form of online services like Google Earth and Yahoo maps, and organizations commonly use data derived from GNSS systems and other sources, such as satellite imagery and photography, aerial photography, and physical surveys, to enhance the kind and quality of online digital information available to them. Government departments, bureaus, counties, community services, utility companies, and school districts are some of the access points for information that may be publicly available. Private companies (e.g., real estate and title companies) may also readily allow access and use of their data on request.

8.3   SYSTEM DESIGN PROCESS

Application requirements, data resources, and personnel are all important in determining the optimum GIS software and hardware solution.

Defining a GIS system for a project is primarily based on the GIS user needs assessment (e.g., as previously shown in Figure 8.2). Some projects are such that there will be no need to worry about the software and hardware design (or selection) if such a system already exists. Other applications will require either the selection or design of a GIS system for meeting the user requirements. In such cases, the design process will generally comprise of: (1) the *physical design*, which includes the hardware and software requirements; (2) *data design*, which includes data representation in the form of entities, attributes, and relations (see, e.g., Section 8.4.2), and design of database and data management layers; and (3) *process design*, which includes data input, update, query, geographical analysis and forms, and other editing functions.

An overview of the methodical approach to a GIS system design will follow shortly. First, let us consider the factors in selecting a GIS system.

8.3.1   SELECTING A GIS SYSTEM

Selecting a GIS system can be a complex and confusing process. It is a critical step, especially when considering buying a new system for a specific application. The system should be researched, selected, tested, and questioned before purchase. Informed choice is the best way to select the best GIS system.

The process of selecting and implementing a new GIS should include: (1) a feasibility study, (2) selection and installation of hardware and software, (3) installation of database, (4) creation of an application program, (5) testing of system with pilot project, and (6) preparation of an application.

GIS systems have three important components: the computer hardware, software, and the organizational context. The computer hardware component considerations will include things like CPU capacity, digitizer, and plotter capabilities. Most important, a GIS software package should match the minimum requirements of the project. If the GIS does not match the requirements, no GIS solution will be forthcoming. The "critical" functional capabilities (with examples in parentheses) are: data capture functions (multiple formats, import and export, digitizing, scanning, and editing), data storage functions (metadata handling, format support, and compression), data management functions (database systems, address matching, layer and management), data retrieval functions (map overlay and locating and selecting by attributes), data analysis functions (statistical analyses and interpolation), and data display functions (desktop mapping, interactive modification, and graphic file export).

A knowledge of the organizational context will ensure that the selected GIS system is not overcapacity. In other words, it should be cost-effective. Other considerations include cost, upgrades, training needs, ease of installation, maintenance, documentation and manuals, and vendor support.

8.3.2 DESIGNING A GIS SYSTEM

The methodical approach to a GIS system design includes a GIS needs assessment and a system architecture design. The system architecture design is based on user workflow requirements identified in the GIS needs assessment. The most effective system design approach considers user needs and system architecture constraints throughout the design process. Figure 8.6 provides an overview of the system design process.

The GIS needs assessment includes a review of user workflow requirements and identifies where GIS applications can improve user productivity. This assessment identifies GIS application and data requirements and an implementation strategy for supporting GIS user needs. The user requirements analysis is a process that should be accomplished by the client or the GIS user organization. A GIS professional consultant familiar with current GIS solutions and customer business practices can help facilitate this planning effort.

The system architecture design is based on user requirements identified by the user needs assessment. Client(s) must have a clear understanding of their GIS application and data requirements. System design specifications can then be developed on that basis. The design and implementation strategies should hardware purchase requirements (if needed) in time to support the user deployment needs.

The system design begins with the technology overview. Client participation is a key ingredient in the design process. The design process includes a review of the existing computer environment, GIS user requirements, and system design alternatives.

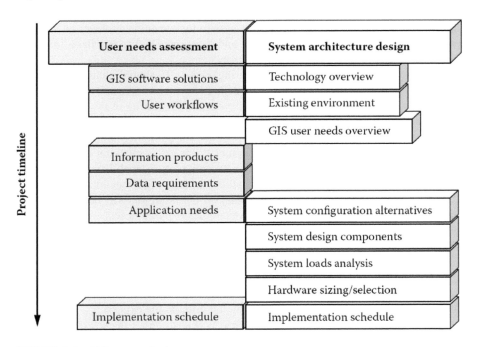

FIGURE 8.6 GIS system design process.

Traditionally, the user needs assessment and the system architecture design can be two separate efforts. There are some key advantages in completing these efforts together. GIS software solutions should include a discussion of architecture options and system deployment strategies for each technology option. The hardware and technology selection should consider configuration options, required platforms, peak system loads for each technology option, and overall system design costs. And finally, the system implementation schedule must consider delivery milestones.

8.4   INPUT DATA FROM FIELDWORK

8.4.1   UTILIZING GNSS FOR FIELDWORK

8.4.1.1   Why Use GNSS?

The use of GNSS technology in the digital mapmaking process has made possible a number of innovations, including the integration of GNSS data into aerial photography expeditions, with exact GNSS positions being recorded at the time of each photographic exposure. These images and coordinate data are then imported into GIS maps. On the ground, portable and lightweight GNSS devices are used to collect positions and attributes of physical geographical features, with the classification of attributes assigned from a pull-down menu. The data can then be output to popular GIS software applications for compilation into digital maps. Thus, GNSS is a valuable resource for GIS applications.

Technological innovations in GNSS and GIS have occurred on a parallel course, with breakthroughs in each field often benefiting the other. The increasing ubiquity of the Internet and the growing affordability of GNSS and GIS systems should lead to increased visibility of these technologies, as seen in the availability of digital maps found on Google Earth and Yahoo.

GNSS allows recording of locations of people, phenomena, buildings, and other objects of interest with minimal effort and minimal cost. It is a much more time- and cost-efficient means of recording positional data compared to alternative methods, such as traditional land surveying. It is also more accurate than approximating the coordinates of places or objects from hardcopy maps. Other notable benefits of using GNSS include its 24-hour everyday worldwide availability, all weather performance, and a uniform global coordinate system for positioning.

Spencer and colleagues (2005) have outlined the following three particular uses of GNSS in the context of a GIS application:

Mapping. The most basic use of GNSS is for mapping. It can be used to update reference maps, or to map objects whose locations were previously unknown or inaccurate. GNSS is most often used to collect coordinates of individual locations, which are realized as *points* in a GIS system. Some GNSS receivers can also be used to collect and generate *line* and *area* (or *polygon*) data. This can be done in a variety of ways. One method is to collect discrete points along the length of a line or at each corner of an area. Another method allows the user to create area data from a single point. A point is collected on the ground within an area of interest (e.g., forest, park, land parcel). The point is then overlaid on top of contextual data, such as an aerial photograph or satellite image, within a GIS. This allows the user to positively identify and outline the area.

Spatial analysis. GNSS is the fastest and most accurate method for providing point location coordinates for performing spatial analyses on mapped data. Social scientists are utilizing GNSS to link sociodemographic survey data to household point locations, and health care data to clinic or hospital locations. Environmental scientists are using GNSS to locate field plots and analyze the characteristics of the vegetation, find animal roosts and examine the spatial determinants of roost locations based on the surrounding landscape, and identify locations of fishing debris in the ocean.

Ground truthing. Ground truthing is the collection of locations and corresponding information about features on the ground that will be used to create, correct, interpret, or assess accuracy. Two common uses of ground truth data are for georeferencing aerial photographs or satellite images. GNSS data collected for ground truthing (or georeferencing) purposes is called geodetic control. The main goal is to collect GNSS control points at locations that are static and easily recognizable in an image, such as road intersections. Then the coordinates can be applied to the pixels of those static features in the image, a transformation algorithm is calculated and applied to the image, and the image is transformed so that every pixel in the new image has accurate coordinates.

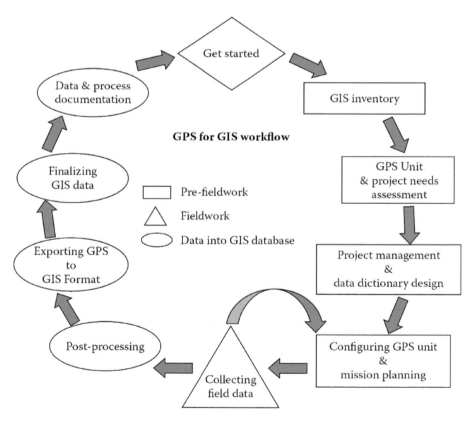

FIGURE 8.7   GPS for GIS workflow.

8.4.1.2 Defining Data Collection Goals and Objectives

Effective use of GNSS requires careful planning before going out to the field. The specifics of the planning will vary depending on the type of project (e.g., types of locations to be collected, and whether they are in a forested area or urban canyon). Figure 8.7 shows a suggested workflow that can be followed formally or informally when using GNSS to collect data for a GIS application.

It is important to know for what objects or features are locations being collected, and why they are important. These questions will relate to the primary goals and objectives of the project. The GNSS coordinates collected must serve their purpose, primarily to meet the project's accuracy needs. For instance, will centimeter-level accuracy or being within 10 and 15 m accuracy suffice for the point locations? The answer to this will have implications on the type of GNSS receiver to be used for the fieldwork.

Once project needs (specifically accuracy requirements) are established, the next step is to design the data collection methods and protocols. This will determine how the GNSS points are observed and recorded. For instance, positions can be recorded at each break or curve in a line, or if a receiver has the capability, it can be turned on and positions continuously recorded while traversing the length of a line or the perimeter of an area.

Given that GNSS data are collected and stored in a digital format, it is easy to add them as geographic features in a GIS. Some GNSS receivers output data in a specific format that requires manipulation before it can be entered into a GIS. However, most receivers that are designed for GIS are capable of providing output data in standard GIS formats. Such receivers will also allow attributes to be added to the data as it is being collected in the field (i.e., the data dictionary concept). This speeds up the fieldwork process and eliminates the need for a separate data entry, import, and linking process after the fieldwork.

Prior to the fieldwork, equipment and training materials should be developed and prepared. The GNSS equipment should be capable of recording coordinates and other pertinent information. Training materials can be used to train the field crew prior to the fieldwork. Such training may include, in addition to the project specifics, what GNSS is and how it works, instructions on the use of GNSS equipment, and basic troubleshooting.

Other mission planning objectives should address logistical and administrative challenges such as accessibility of features to be mapped, health and safety precautions, scheduling of data collection, and planning for meetings for data download and processing.

8.4.2 BASIC DATA DICTIONARY CONCEPTS

A data dictionary is a list of features and attributes to be collected for a particular GIS application. Using a data dictionary enables point, line, and area features to be created from the GNSS coordinates collected in the field. Although the amount of detail to be captured depends on the purpose of project, proper planning should ensure that only the right amount of information is captured for that project (Figure 8.8).

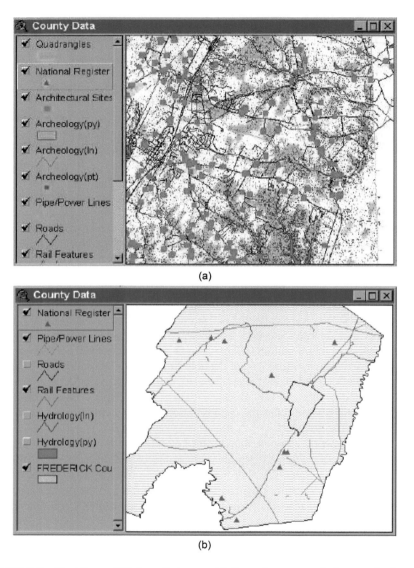

FIGURE 8.8 Deciding a GIS data dictionary. (a) Capturing too much data results in complex maps, difficult to read, (b) Capturing too little data results in incomplete maps and faulty analysis.

Capturing too much information may lead to unnecessary time and cost, and complex maps that are difficult to read. On the other hand, capturing too little data may lead to faulty analysis.

Developing a data dictionary: Some of the GNSS technology that have been developed for GIS applications enables the creation of a "job" or a "project" such that all files in a project should use the same data dictionary. This ensures smooth processing between the GNSS software and the GIS application. The first step is to

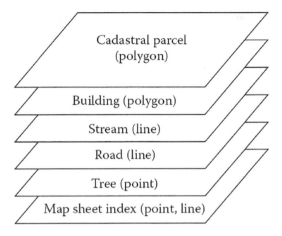

Cadastral parcel
(polygon)

Building (polygon)

Stream (line)

Road (line)

Tree (point)

Map sheet index (point, line)

FIGURE 8.9   Feature classification layers.

identify the features to be observed and mapped, and to determine the attributes for each feature.

Feature classification: All features must be classified as either a point, line, or area (polygon) (Figure 8.9). Feature classification depends on the use of the data and the scale at which it will be displayed (Figure 8.10). In that process, it is also important to identify both the target features and reference features that provide context and quality control (Figure 8.11). An equally important task is to determine the attributes to record—collect only attributes that are recognizable in the field (e.g., roof type, construction material, architectural detail, etc.).

8.4.3   OTHER METHODS

GNSS methods for collecting GIS data are generally faster and accurate. Traditional ground survey methods can be more time-consuming and expensive, and sometimes, problems and obstacles associated with implementing them can lead to low accuracy in coordinates. Still, there are (and have been) instances where other methods can be used to incorporate spatial perspective into a GIS. Four such methods include approximating coordinates from hardcopy maps, traditional surveying, digitizing, and address geocoding (Spencer et al., 2005).

8.4.3.1   Using Maps to Approximate Coordinates

Using maps to approximate coordinates method is the oldest and has been used for centuries. It is based upon a map having a well-defined coordinate system grid (or graticules). The grid system is common to all topographic maps and is basically a crisscross of horizontal and vertical lines that have well-defined coordinates, separated by a regular interval. There are two ways to approximate the coordinates of a given feature on such a map.

The simplest way (but not highly accurate) is to eyeball the location. In other words, find the feature on the map and approximate its coordinates using the

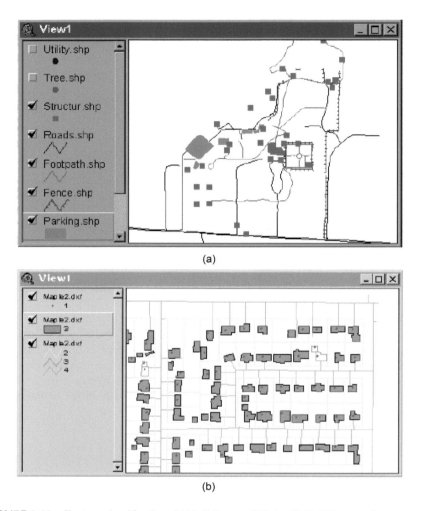

FIGURE 8.10   Feature classification. (a) Buildings as Points, (b) Buildings as Areas.

coordinate grid system. This method is the quickest and only gives a general idea of the feature location coordinates.

The second (more accurate) way is to approximate coordinates from the map using a ruler or protractor. This method requires that the measured map distances will be based on a common map scale. The distances are measured as offsets from the horizontal and vertical grid lines nearest to the feature. However, even with the most detailed map ruler, accuracy will depend on the map scale (see Table 8.2). Other possible error sources include human error and condition of the map (map wrinkles can skew distance measurements).

With either of these two methods, each feature has to be measured and coordinates recorded for use into the GIS system. This can be an extremely tedious and time-consuming process depending on the size of the project.

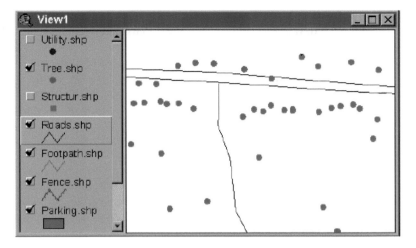

FIGURE 8.11 Feature classification. Reference features provide context and quality control. (Trees are the target features; roads are the reference features.)

TABLE 8.2
Scale-Dependent Errors

Map Scale	Expected Error (m)
1:1,000,000	508.0
1:250,000	127.0
1:100,000	50.8
1:25,000	12.7
1:10,000	8.47
1:5,000	4.23
1:1,000	0.85

Source: Based on U.S. National Map Accuracy Standards.

8.4.3.2 Traditional Surveying

Traditional surveying methods are well-established methods for locating position coordinates from trigonometric calculations using measured angles and distances. They include, for example, use of directional theodolites, EDM or total stations, levels, and terrestrial laser scanners. The main advantage of such methods is that the resultant coordinates are extremely precise; however, that advantage is countered by the high costs of equipment and labor, and the longer time required to complete the fieldwork. In addition, these methods are affected by weather and environmental conditions on the ground. GNSS, on the other hand, is an all-weather technology and is available anywhere. Traditional surveying would be good if there is an established reference (control) network, proper equipment is available, and only a few points need to be coordinated. It would not be efficient for a large number of points.

Most of the traditional surveying methods would still require that field data are manually recorded, processed, and manually entered into the GIS database. GNSS systems, on the other hand, have been developed with the capability to provide data in digital format that can be readily imported into a GIS. Some GNSS equipment can be less accurate than traditional methods, depending on the type of equipment and corrections applied to the GNSS data.

8.4.3.3   Digitizing Coordinates

Digitizing hardcopy maps is another method for obtaining GIS input data. In this method, a hardcopy map is placed on a large flat tablet that is connected to a computer, and then points and lines are digitized using an electronic sensor with crosshairs. Each location is entered into the GIS with more accurate coordinates than can be obtained using the manual methods.

Digitizing coordinates allows for greater accuracy than a map ruler and substantially reduces the user error. It is also faster for collecting feature information, although the high cost of hardware and software can be prohibitive. In addition, the data acquired by this method are already in digital format and there is no need to preprocess or manually enter them into a GIS.

The only other error sources would be the digitizing coordinate transformation, map quality, map scale (see Table 8.2), and a new kind of user error—the ability of the user to accurately trace the points and lines being digitized.

8.4.3.4   Address Geocoding

Address geocoding is a recently developed methodology for locating people and places. Instead of using hardcopy maps or equipment, it is based on use of roads and addresses within a GIS to obtain relative locations for people and places. The method requires a data set containing a road network with road names and address ranges for every road segment. It may be limited to applications locating people and places because other features that are not closely aligned with a road network may not be located in this manner. In reality, many areas of the world do not have properly established road networks with proper naming and addresses, precluding the use of this method.

8.5   OTHER SOURCES OF GIS DATA

Although in many instances GIS inputs involve use of traditional maps and field data collection workflows, various other sources exist. These include, for example, satellite remote sensing, Earth Observation services, and Big Data. Remotely sensed data are used as a background or as inputs to make informational maps, conduct spatial analyses, and perform geoprocessing functions within a GIS software. Satellite images and Big Data form the basis for GIS analyses for many applications, for example, augmented reality (AR) applications, environmental monitoring, transportation planning, city planning, troop movements, and so forth.

8.5.1 Augmented Reality Informatics

An AR technology overlays digital information such as a virtual 3D model (of buried pipes and underground utilities, a building design, a road design, earthwork volumes, etc.) on a live image of what is being viewed through a device such as a smartphone camera (Ogaja 2022). The intent of AR is to overlay the virtual 3D model either as a copy of the real object as it exists in situ, or as a design of something to be added or built, at the location being viewed through the camera display screen of the device.

Primarily, an AR system comprises a mobile device, handheld or head-mounted, with an integrated display screen, a camera, processor, a Global Positioning System (GPS) or GNSS, inertial orientation sensing, microphone, and other components. Because devices such as smartphones and tablets already have such hardware, AR systems have been developed that makes use of the same as shown in Figure 8.12.

AR systems enable users to carry out many useful functions such as:

- Onsite visualization and measurement of existing features
- Onsite verification of existing GIS data
- Automatically placing models on site with high (centimeter-level) accuracy
- Viewing subsurface virtual models of underground utilities
- Visualize, measure, and record features in the field with ease
- View attribute-rich models live in 3D

8.5.2 Remote Sensing and Earth Observation Services

Data from drones, piloted aircraft, and earth orbiting satellites are important and commonly used sources for GIS applications. Such data provide remotely sensed imagery and terrain models for mapping and image analysis. The end products include maps, reports, and web services such as Google Earth services, Google Maps, Digital Earth Africa, Digital Earth Australia, Nearmap, and many other derivative products and services. *Airborne data* and *imagery, satellite imagery*, and maps are the vital components of modern GIS systems. Such data are often generated by integrated systems (e.g., drones and/or aircraft with GPS/GNSS and imaging sensors, satellites with precisely determined orbits, etc.) and are therefore much more precisely georeferenced for GIS applications.

8.5.3 AI and Big Data

Artificial Intelligence (AI): Computer scientist, John McCarthy, coined the phrase in 1956 and described it as the science and engineering of making intelligent machines, especially intelligent computer programs. In the field of computer science, AI refers to the use of computers to do tasks that are usually done by humans—mimicking human intelligence. However, these are not just any tasks usually done by humans, but the tasks that require use of human intelligence and discernment. As part of AI, *Machine Learning* (ML) is defined by Oxford dictionary as *the use and development of computer systems that are able to learn and adapt without following explicit*

(a)

(b)   (c)

FIGURE 8.12 Screenshots of augmented reality display systems. (Images used by permission. Courtesy of Trimble Inc. (a) and Augview Ltd. (b & c).)

instructions, by using algorithms and statistical models to analyze and draw inferences from patterns in data. This then brings us to the definition of "Big Data":

> extremely large data sets that may be analyzed computationally to reveal patterns, trends, and associations, especially relating to human behavior and interactions.
>
> *– Oxford Dictionary.*

ML is therefore a subset of AI. ML applies computer algorithms to build data-driven models that improve automatically through self-learning to make predictions and decisions, guided by training data and experience, all this taking place without any explicit programming. From recent technological developments, the following have also become part of ML in common data-driven discourse: deep learning, probabilistic learning, reinforcement learning, transfer learning, decision trees, and genetic algorithms. Hence, there is a connection between *AI* (and/or *ML*) and *Big Data*.

Big Data therefore simply means extremely large data sets which reveal key insights when analyzed computationally. It has been discussed in four key dimensions of interest, usually being conceptualized as 4Vs: *Volume, Variety, Velocity,* and *Veracity.* In Big Data, actionable location-based intelligence is critical to decision support and is a key requirement for GIS analysis, in this respect pointing to the important role of GNSS and other positioning technologies that aid in generating coordinates to assist in geocoding.

But what is the role of AI and Big Data in a modern GIS?

The rapid increase in data volumes, mainly semi-structured and unstructured data that are generated as a mix of text, pictures, audio, and video files, necessitates the application of AI/ML for faster and more accurate processing, analysis, and modeling for decision support. AI/ML enables increased speeds, efficiencies, and accuracies in GIS analysis and spatial modeling, hence the preference for ML algorithms, such as Support Vector Machine and Random Forest as opposed to traditional classification methods, such as the maximum likelihood classifier in ArcGIS.

> AI is a data-driven game.
>
> *– Sud Menon, Esri; and GIS provides location intelligence*

In other words, GIS uses Big Data to provide insights into location information in a way that informs *smarter* decision-making. For example, in a megacity, AI/ML can be used to predict or reveal geospatial patterns of different data types such as on pollution rates, traffic density, crime rates, geohazards such as earthquakes and flood risks, employment, population growth and distribution, housing demand, and health care needs, based on prior historical records and pertinent information. The Big Data from AI predictions are an output that can also serve as input for other subsequent GIS analyses and/or location intelligence that can serve in smarter planning and logistics workflows.

BIBLIOGRAPHY

Coleman, D. 2007. *Manage Your GeoProject Effectively: A Step-by-Step Guide to Geomatics Project Management.* Calgary, Alberta, CA: EO Services.

ESRI Inc. 2009. *System Design Strategies.* 26th ed. Redlands, CA: Environmental Systems Research Institute, Inc.

Ghilani, C. D. and P. R. Wolf. 2008. *Elementary Surveying: An Introduction to Geomatics.* 12th ed. Upper Saddle River, NJ: Prentice Hall.

Ogaja, C. 2022. Augmented reality: A GNSS use case. In: *Introduction to GNSS Geodesy.* Cham: Springer. https://doi.org/10.1007/978-3-030-91821-7_1.

Spencer, J., Frizzelle, B. G., Page, P. H., and J. B. Vogler. 2005. *Global Positioning System: A Field Guide for the Social Sciences*. 3rd ed. Malden, MA: Blackwell Publishing.
Trimble Inc. 2022. *Trimble Sitevision — High Accuracy Augmented Reality System*. https://sitevision.trimble.com (accessed August 25, 2022).

EXERCISES

1. What is a GIS application? What are the steps in developing a GIS application?
2. Which one of the following is/are not GIS application(s)?
   a. Design of a wireless GNSS receiver for extreme terrain conditions
   b. Selection and implementation of a new GIS in a utility company
   c. Redesign of a geospatial imagery database
   d. Production of a regional wildlife habitat map
   e. Creation of an environmental database for a nuclear power plant
   f. Development of a national land information system
   g. Development and implementation of a Web-based national biodiversity mapping service
   h. Mapping of route options for a proposed pipeline
   i. Surveying a new subdivision
3. The following is an excerpt from the U.S. National Map Accuracy Standards:
   Horizontal accuracy. For maps on publication scales larger than 1:20,000, not more than 10% of the points tested shall be in error by more than 1/30 inch, measured on the publication scale; for maps on publication scales of 1:20,000 or smaller, 1/50.
   a. What will be the scale-dependent error in horizontal coordinates obtained from a map of scale 1:75,000, assuming that such a map has a note in the legend saying, "This map complies with National Map Accuracy Standards"?
   b. Supposing you can roughly estimate the location coordinates of a feature from a 1:500 topographic map, would this be sufficient for a GIS application requiring 1 m accuracy?
4. Explain what is meant by *georeferencing* and why GNSS is the preferred method for GIS georeferencing.
5. There are a number of operational GIS systems in professional and educational environments. Examples include ArcGIS (by Esri), ArcPAD (mobile GIS designed for GPS and PDA), AutoCAD Map (by AutoDesk Inc.), GRASS (a UNIX GIS developed by the U.S. Army Corps of Engineers), IDRISI (open code developed at Clark University), Maptitude (Caliper Corporation), GeoMedia (CAD software with GIS extensions, Intergraph Corp.), and MapInfo (uses Visual Basic, favored for 911 and other applications). Select any two systems, either from this list or any other source, and compare them in terms of the following:
   a. Operating system/platform
   b. Targeted applications
   c. Advantages and disadvantages

9 Unmanned Aerial Systems Surveys/ Mapping

9.1 INTRODUCTION

Unmanned aerial vehicles (UAVs), also variously described by other terms such as unmanned (or uncrewed) aerial systems (UASs), are commonly used by both recreational and professional users. In geomatics, these tools have become commonplace in aerial surveying and mapping projects by licensed practicing professionals. This chapter presents important considerations when planning or designing an aerial surveying/mapping project with a UAV. But first, let's include a few definitions to avoid confusions.

The term *drone* refers to any vehicle that is controlled remotely in any medium such as land, air, ocean, sea, and lake. It is a general term that describes all autonomous vehicles including, but not limited to, the ones used in the military.

UAV is an aircraft with no pilot on board, whereas a *UAS* is a system, of which the components include aerial vehicle and associated equipment that do not carry a human operator, but instead fly autonomously or are remotely piloted. Thus, the *UASs* include the command/control/communications systems and the personnel necessary to control the unmanned aircraft.

A key distinction between the terms UAV and UAS is that a UAV is the actual aircraft that flies autonomously, while the UAS comprises the UAV, remote pilot, remote controller, radio link, batteries, and propellers, among other accessories.

In UAS surveys, the UAV is usually equipped with a camera sensor (payload) which captures images in a fluid motion of a flight plan. Each image captured is assigned specific georeferenced coordinates by the UAV's positioning and navigation system. The images are captured with a certain percentage of overlap which enables the construction of a 3D image. The main project goal is to plan for research and preparation, data collection, analysis, and presentation.

9.2 GENERAL CONSIDERATIONS

9.2.1 UAS LAWS

It is important to note that UASs are not toys and so not everyone is supposed to use them. Their use is closely governed by the Civil Aviation Authorities of respective countries. The United States, for example, requires a person to hold a part 107 FAA license to operate a UAV commercially. The term "commercially" can sometimes be used loosely but it means using the UAV to generate income such as, for example, making YouTube videos, real estate photography, survey and mapping, and wedding photography.

Similarly, other countries have their pertinent laws and regulations. For example, in Kenya, an East African country, a person needs to hold a drone license to operate a UAV and the drone must be registered and possess markings that follow the country's plane system (such as the 5Y-001A plane marking system). Additional requirements include such as being in possession of a Remote Operator Certificate and a permit for flying the UAV for the day.

Although the laws and regulations vary from country to country, there are some similarities which are guided by the need to provide both private and public safety, protect life and property, and avoid infringement of privacy rights, among others. Examples are as follows:

1. Not flying above 400 ft (120 m) above ground level and within 50 m of any person, vehicle, or structure which is not under control of the drone operator. There are exceptions to this rule provided prior permission is sought from the regulator.
2. Not flying in conditions other than Visual Meteorological Conditions.
3. Not flying at night except where expressly exempted by the regulator.
4. If drones are fitted with cameras, not shooting photos or videos beyond the prescribed area of approved operation.
5. If camera-fitted drones shoot photos or videos beyond the specifications above, then the same should not be distributed or published.
6. Reporting accidents or incidents involving drones to the regulator.
7. Notifying the regulator in case of loss or theft of the aircraft.
8. Not flying a drone over a public road or along the length of a public road at a distance of less than 50 m, except where exempted.
9. Not operating within 10 km of an airport, or on Approach and Take-off paths.
10. Not operating a drone that has not been insured against third-party risks as a minimum.
11. Not using a drone to conduct surveillance of a person or of private property without the consent of the person or owner, respectively.

As noted, the regulations vary from country to country with some having sketchy regulations to just give guidance, while others have elaborate laws complete with fines for non-compliance. It is therefore important to research and become familiar with the local laws and regulations which are pertinent in the locality of your project.

9.2.2 PROJECT SCALE AND LOCATION

The first step to any drone survey project is to determine the project scale and location. This will help the drone surveyor, for example, to determine which drone will be appropriate for the job, the number of days to be spent on the field, whether you need clearance from the air regulator to fly in the area of interest, the number of ground control points (GCPs) needed, any obstacles to note in the project's area of interest, and so forth.

It is important to check the local weather forecasts (predictions) for the project's location. This is because certain weather conditions are not suitable for UAV flights, such as winds of more than 10 m/s and visibility of less than 10 km. Many tools and applications have been developed by vendors to help in assessing the weather in advance, especially for UAV projects. These include, for example, the UAV Forecast app (www.uavforecast.com) and the Windy app as shown by screenshots in Figure 9.1a and b. Applications such as AirMap (Figure 9.2) can also be used to check the airspace of the project site to determine if there is a no-fly zone for drones.

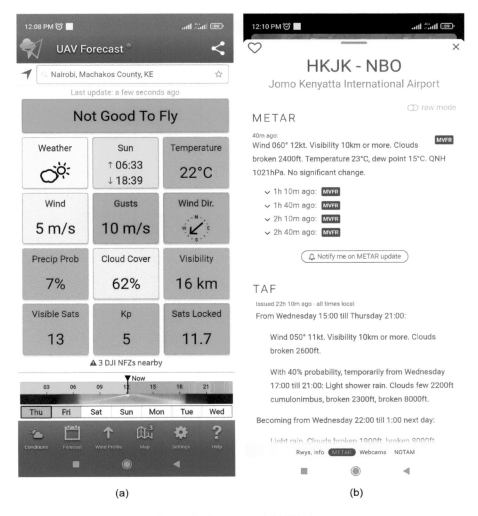

(a)　　　　　　(b)

FIGURE 9.1 Screenshots of (a) UAV forecast and (b) Windy app.

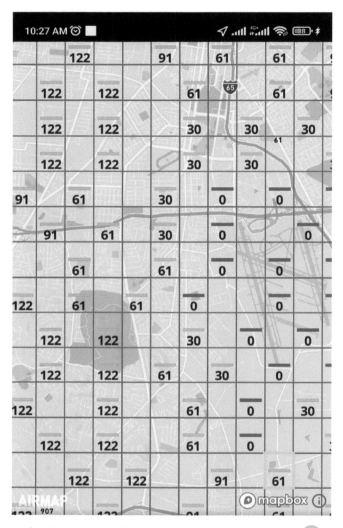

FIGURE 9.2 Screenshot of AirMap showing restrictions of flying in Louisville, Kentucky.

9.2.3 MISSION PLANNING AND FLIGHT CONSIDERATIONS

Type of UAV/payload: Depending on the kind of job to be undertaken, no two UAVs are alike. Generally, all the UAVs fall into two major categories: fixed-wing and multi-rotor types.

Fixed-wing UAVs function like a jumbo jet (see example in Figure 9.3). They have wings and get their forward movement from a single tail propeller. They tend to have longer battery life and hence are better suited for job sites with large acreage. Their main drawback is that they require a clear take-off and landing site.

Multi-rotor UAVs on the other hand take-off like a helicopter, in an upward motion (see an example shown in Figure 9.4). They pitch forward and backward, bank left and right by tilting the airframe and by varying the speeds of the individual rotors. They can also rotate in one position in a movement referred to as "yawing."

Multi-rotor UAVs are quite flexible in movement, but they have shorter flying times because of the energy required to power the fast rotations of the brushless motor. This means that when deployed for missions, the users must carry extra batteries or a source of power to recharge them once they are out.

Technological advancements have led to the development of another type of UAV which is a hybrid of fixed-wing and multi-rotor. This UAV takes off and lands like a multi-rotor but when it reaches the desired flying altitude, it flies like a fixed-wing UAV. It can overcome the challenges of a clear take-off and landing site faced by fixed-wing UAVs and the quick drainage of battery life by the multi-rotor types. It presents the best of both worlds and would be the most desirable to have for projects. Figure 9.5 shows an example image of a hybrid UAV.

With regards to payload, the factors that one should consider when choosing the UAV include camera resolution, ease of operation, and storage memory of the UAV. A good resolution to work with is 20 MP on the minimum.

Flight planning: This is a very important step in mission planning. Once a UAV or drone type and its payload have been selected for achieving desired results, the next

FIGURE 9.3 Image showing a fixed-wing UAV.

FIGURE 9.4 Image showing a multi-rotor UAV.

FIGURE 9.5 Image showing a hybrid UAV.

step is to determine the parameters it will use during the flight. These parameters can make or break the survey project. Therefore, it is important to pay close attention during this early stage.

In most cases, flight parameters are usually defined in a drone flying application. Vendors such as DJI GO4, Flylitchi, and DroneDeploy provide such applications for flight planning. Some of these applications have desktop versions that allow one to define the parameters carefully in the office and the same can be saved and opened on mobile versions. This ensures that mistakes are minimized as much as possible and to avoid hurried set ups on site.

Examples of flight parameters to be defined include the flying height, percentage of overlap (forward and side overlap), camera position, UAV speed, and type of flight. These are briefly described as follows:

- *Flying height:* This parameter largely depends on the use case of the final product (i.e., the intended application or use of the final orthomosaic from the aerial survey). *For drone surveying projects, the height should not go beyond 80 m above ground level; otherwise the accuracy of the orthomosaic product will be affected and its use in design will be untenable.* The flying height affects the Ground Sampling Distance (GSD) which is defined as the actual length (distance) on ground which is represented by one pixel on a digital map. Therefore, if the GSD of a map is 5 cm, it means that one pixel represents 5 cm on the ground. The higher the flying height, the higher the GSD and vice versa. Drone surveying projects demand a low GSD of between 1 and 2 cm. This can only be achieved by flying at a maximum of 80 m above ground level.
- *Percentage of overlap:* This is an important concept in photogrammetry and is well covered in many textbooks. Overlap *is the amount by which one photograph includes the area covered by another photograph and is usually expressed as a percentage.* The distortions and faults caused by the camera lens, tilt, and relief displacement are more at the edges of the photograph. If the overlap is more than 50% these distortions and faults can be eliminated. The overlap portion of pairs of photographs is useful for stereoscopical 3D viewing of the area covered.
  - For redundancy and due to overlap requirements, each portion of a particular area is photographed a few times (e.g., four to five times). This ensures that any photograph with distortion, faults, or cloud shadows can be rejected without the necessity of taking new photographs.
  - If the UAV flight lines are not maintained straight and parallel, there would be gaps between adjacent strips. These can be avoided by having side overlaps.
  - It is also important to consider what the optimum forward and side overlap of a good UAV flight would be. Generally, 75% overlap is considered optimum for both overlaps. But why not have more overlap, like 95%?
  - Studies have shown that an overlap of more than 80% does not come with a commensurate increase in accuracy. Instead, it introduces a drawback, that is, it increases the number of photographs and subsequently more time spent during the software processing phase.
- *Camera position:* The camera needs to be pointed to the nadir position (−90°). Pointing to the nadir means that the camera axis (in the direction of the lens) is perpendicular to the ground. The images resulting from this position are known as vertical images. When the camera is not pointing perpendicular to the ground, the resulting image is known as an oblique image. In UAV surveying, the practice is nadir photography.
- *UAV speed/camera interval:* A carefully calculated flight speed will help the drone fly fast enough to be efficient but also slow enough so that the data

collected are of high quality. When the goal is high-quality data, which is the case in UAV surveying, the practice is usually to fly slower. The speed is a factor determined by the image interval (defined by the amount of overlap in your images) and the camera's photo interval, which is the time taken by the camera to write the image onto an onboard computer memory unit such as the SD card. Optimal flight speed = image interval (m)/camera interval (s).

- *Type of flight:* There are two types of flights in mission planning: grid flights and waypoint flights. The two modes are automated flight modes available on many UAV mission planning applications. Each mode enables a pilot to plan and execute a choreographed flight or set of flights, ensuring repeatable and standardized data collection in the field. Grid flights are displayed by grid lines. Waypoint flights are displayed as connected dots.

  - *Grid flights* are generated automatically from the mission planning application once the user has defined the area of interest. The number of grid lines the application generates is a factor of the overlap percentages. The higher the overlap percentage, the higher the number of grid lines and vice versa.

  - In *Waypoint flights*, the user selects the UAV flight path by defining a series of pit-stops and what action the drone will take once it gets to that pit-stop or waypoint. For instance, the user can define that the UAV makes a rise of 20 m in between two pit-stops and make a drop of 25 m in the next two pit-stops. In UAV surveying, waypoint flights are preferred in hilly areas where there is a high risk of the UAV crashing on the side of a hill because of sharp changes in the terrain. Grid flights are preferred when mapping relatively flat areas.

9.2.4 Ground Control Points

A GCP is an actual physical point on the Earth's surface with a known geolocation or position coordinates (see Figure 9.6). Defining a project's GCPs is an important step in georeferencing the aerial mapping/survey project.

Georeferencing is the process of taking a digital image and adding geographic information to the image so that GIS or mapping software can 'place' the image in its appropriate real-world location (or rather a ground system of geographic coordinates) – Science Education Resource Center at Carleton College

GCPs can be constructed in many ways, for example, using white bucket lids and black tape. However, whichever method is used, a GCP needs to be big and clear enough to be detected by a UAV camera lens. For example, the sharp contrast of a black tape with a white bucket lid ensures that it can be detected by the camera lens hundreds of feet above.

As illustrated in Figure 9.6, measurement of GCPs is typically done by Global Positioning System (GPS)/GNSS receivers which are capable of centimeter-level georeferenced accuracy. The georeferenced GCPs are vital for ensuring that the final products of the UAV surveying can be applicable in subsequent use-cases such as in infrastructure design or planning.

FIGURE 9.6 Photo showing an example of a ground control point (GCP) measurement with a GPS/GNSS receiver.

The number of GCPs for a particular acreage is also an important factor. The goal is to ensure that every control point in the final orthomosaic product will contain geographic coordinates, and that there are enough GCPs to provide good aerial triangulation in the processing software. The distribution of the control points should also be done equitably so that the processing software can deliver an accurate final orthomosaic. For illustrative purposes:

For sites below 24 acres, there should be five GCPs to balance the site: four GCPs on the furthest corners (assuming a rectangular site) and one in the middle. For sites above 24 acres, the following formula ought to be used in determining the number of GCPs.

No. of GCPs = Acres × 0.2

This translates to one GCP for every 5 acres of land. Again, to re-emphasize, this is for sites above 24 acres. When GCPs are calculated this way, the final product gives very good XYZ RMSE tolerances.

9.2.5 POST-FLIGHT PROCESSING

After completion of all the field work, the captured images are downloaded from the UAV's computer storage (such as SD card) and imported into the processing software. There are several UAV image processing software applications by different vendors, offering different features and capabilities. A few examples include Pix4D, Agisoft Metashape Pro, Carlson Photo Capture, and DJI TERRA, among others. Due diligence is necessary to ensure that the right software is used to provide an efficient solution for a project's workflow.

Once the images are imported into the processing software, depending on the software of choice, the user can view the flight lines and image spacing against a background map of the site. A next step would be to align the images, bring in the ground control file, tag the GCP targets, and confirm that the XYZ RMSE values are within acceptable tolerances. Mostly with proper training in the use of such software applications, the processing is intuitive and automated steps make it easy to do functions such as flagging of outliers.

Subsequent steps would include building the dense point cloud, the digital elevation model (DEM) model, and finally the orthomosaic. The computer takes over this process and will complete after several hours. Slower computers can even take a week to process, so the user should ensure that the computer speed meets the software requirements for faster results.

9.2.6 DELIVERABLES

The following are some of the common deliverables from a typical UAV survey project:

- *Dense point clouds:* A point cloud is the representation of a geographical area, terrain, building, or feature which is compiled through a huge collection of points, usually millions or billions, and plotted in 3D space. The virtual representations of the real world can be used to analyze elevations, distances, volumes, and on-site progress to drive understanding, collaboration, and decision-making.
- *DEM:* A representation of the bare ground (Earth) topographic surface of the Earth excluding trees, buildings, and any other surface objects. They are used for 3D surface visualization and for generating contours.
- *Orthomosaic:* An orthomosaic, as the name suggests, is a mosaic of photogrammetrically orthorectified images. They have been corrected for geometric and lens distortions and seamlessly stitched together to form one continuous image.
- *Contours:* Contours are continuous lines that close in on themselves, joining points of equal elevation. They are usually generated from DEMs using specialist software. Before the advent of software-assisted mapping, cartographers used to draw contours using interpolation techniques based on a grid reference with a series of spot heights acting as guidelines for elevation differences. They are required for infrastructural planning and are a necessary prerequisite for engineering designs.

9.3 STANDARD UAS OPERATING PROCEDURES

9.3.1 PRE-FLIGHT OPERATIONS

Activities before the start of a flight operation include inspection of the UAS and any additional equipment, assessment of the operating location, and briefing any project members involved in the operation. All flight operations should be conducted in accordance with the pertinent laws, state and local regulations, and the operator's manual for the UAS.

- The operator and crew should be familiarized with all available information pertaining to the flight such as take-off/landing, including but not limited to any operational limitations due to weather conditions, hazards, no-fly zones, and so forth.
- The operator should be aware of all surroundings in the event that an emergency landing is necessary. This includes the ability (and plans) to recover the UAS.
- Before the first flight of the day, verify all batteries are fully charged.
- Inspect the UAS and related equipment for any signs of damage and overall condition; ensure that camera(s) and mounting systems are secure and operational; and check the entire system per the pre-flight inspection instructions in the manual to make sure it is in good structural condition and no parts are damaged, loose, or missing.
- Repair or replace any part found to be unsuitable to fly during the pre-flight procedures prior to takeoff.
- Perform an overall visual check of the UAS prior to arming any power systems.
- Gather enough information about existing and anticipated near-term weather conditions at the mission environment. It is best practice to use government-approved resources.
- Wind condition is a major factor in flight operations. Precautions should be taken to ensure that wind conditions do not exceed the UAS limits (e.g., as stated in the operations manual). Pocket anemometers are available from a variety of sources and simple to use to estimate the wind speed to determine if it is within the limits of the UAS being flown.
- Go through a checklist (from the UAS vendor, if any). In case that is not available, a standard flight checklist (e.g., Figure 9.7) should be made and followed by the project crew. The operator should utilize the checklist to ensure the highest level of safety. At a minimum, this pre-flight checklist should contain the following:
    - Required documentation such as operator's permit and/or license, UAS operation manual, registrations, insurance, and so forth.
    - Weather conditions.
    - Batteries fully charged and mounted.
    - Communications (datalink) check.
    - Ensure the GPS/GNSS positioning module has a fix.
    - Check mission flight plan.
    - Takeoff and landing locations, including for emergency and recovery actions.
    - Sensor calibrations.
    - Project's crew briefings.

FLIGHT CHECKLIST		
PRE FLIGHT	**DURING FLIGHT**	**POST FLIGHT**
At Office	**After Launch**	**After Landing**
☐ Documentation	☐ Safe altitude reached	☐ Power down UAS
☐ Local regulations & permissions	☐ Confirm observer has UAS in sight	☐ Remove and safely store batteries
☐ Weather permits flying	☐ All systems green	☐ Check sensor/camera to ensure data collected
☐ Batteries charged	☐ Satellite GPS check	☐ Transfer data
☐ Flight Gear check	☐ Check battery remaining	☐ Make logbook entry
In the field	**Before Landing**	**Back at Office**
☐ Scan area for obstacles, e.g., take-off & landing area	☐ Ensure flight has been done according to mission plan	☐ Flight report and UAS maintenance reports
☐ Wind check	☐ Wind check	☐ Charge batteries
☐ Assemble UAS, ensure screws are tight & propeller check	☐ Scan landing area for obstacles	☐ SD card (or data storage device) secured and ready
☐ Sensor/camera check	☐ All systems green	☐ Data processed
☐ Batteries securely mounted		
☐ Ensure GPS fix		
☐ Confirm flight plan		
☐ UAS inspection		
☐ Crew briefing		
☐ Wind check again for launch		

FIGURE 9.7   Example of a UAS flight checklist.

- Ensure launch site is free of obstacles.
- Recheck wind direction before launch.
- Contact information of nearest (or local) Air Traffic Control facility in the event of an emergency.

9.3.2 During Flight Operations

- The UAS operator should launch, operate, and recover from preset locations so that the aircraft will fly according to the mission plan.

- After the UAS is launched, the operator (and crew) should have a clear view of the aircraft at all times, a visual line of sight.
- UAS should not be flown over persons not directly involved in the operations. Populated areas, heavily trafficked roads, and open-air assembly of people should be avoided except only if permission (or waiver) is granted for the project needs.
- An observer should make the operator aware of any hazards during the flight operations.
- In case of any failure during the flight or loss of visual contact with the UAS, the operator should utilize the built-in failsafe features to recover the aircraft and to follow any emergency procedures defined in the operator's manual.

9.3.3   POST-FLIGHT OPERATIONS

- The UAS operator should scan the landing area for potential obstruction hazards and recheck weather conditions.
- Announce to crew and/or people around that the UAS is inbound to land. Carefully land the aircraft away from any obstructions and people.
- After landing:
  - Shut down the UAS and disconnect the batteries.
  - Power down the camera or sensors.
  - Visually check aircraft for signs of damage and/or excessive wear.
  - Verify that mission objectives have been met.
  - If imagery or other data are recoded onboard the aircraft during flight, transfer the data as necessary to a backup storage device. If all data and imagery were transmitted (and recorded) to a ground control station during the flight, then consider double-checking (and backing up) the data prior to conducting additional flight operations, if any.
  - In case there are multiple flights to be conducted, repeat checklist steps to prepare the aircraft for launch again.

9.3.4   EMERGENCY PROCEDURES

- UAS emergency procedures are usually specific to the UAS type and are provided by the manufacturer. It is the responsibility of the flight crew to be familiar with the operational manual provided by the vendor before any flight operations are conducted.
- It is also a recommended safe practice to prepare a checklist for emergencies.
- Some possible emergencies due to system failures include
  - Loss of communications link with the UAS
  - Loss of UAS's GPS/GNSS functionality
  - Loss of engine power
  - Autopilot software failure
  - Intrusion of another aircraft into the UAS mission airspace.

- The operator and crew should be familiar with the UAS operator manual for failsafe options and mechanisms which should also be tested during training and mission preparations, as well as crew briefings.
- Emergency procedures include, for example, landing immediately, if possible, automated return to land, moving to a predetermined location or altitude, and other risk mitigation actions such as immediate incident reporting with pertinent authorities.

9.3.5 INCIDENT REPORTING

- In the event of a lost link with the UAS and/or fly away, the operator should evaluate the airspace affected and contact the appropriate controlling agency. For instance, this might include the local control tower, airport manager, or any other responsible agency in the project's airspace (e.g., the FAA in the United States).
- Reporting the loss of link and/or fly away incident should include details of the flight such as the location, direction of flight, approximate altitude, speed, and the remaining flight time (remaining battery life).
- For projects in the United States, in the event of an emergency the operator should be prepared to submit a written statement if requested by the administrator (FAA) as outlined in UAS operating procedures.

BIBLIOGRAPHY

Crume, J., C. Crume and B. Crume. 2021. *Survey Mapping Made Simple: Workflow (Field to Finish)*. Independently published. ISBN 979-8-5011-6999-9.

Eskandari, R., M. Mahdianpari, F. Mohammadimanesh, B. Salehi, B. Brisco and S. Homayouni. 2020. Meta-analysis of unmanned aerial vehicle (UAV) imagery for agro-environmental monitoring using machine learning and statistical models. *Remote Sensing*. 12. doi:10.3390/rs12213511.

Federal Aviation Administration (FAA). 2016. Operation and certification of small unmanned aircraft systems. *Federal Register*, 81(124): 42063–42214 (152 pages).

Gupta, S., G., Mangesh and J. Pradip. 2013. Review of unmanned aircraft system (UAS). *International Journal of Advanced Research in Computer Engineering & Technology*. 9. doi: 10.2139/ssrn.3451039.

Heliguy. 2022. *Drone Surveying: A guide to Point Clouds*. https://www.heliguy.com/blogs/posts/drove-surveying-guide-to-point-clouds (accessed August 15, 2022).

North Carolina Department of Transportation (NCDOT). *UAS Standard Operating Procedures. NCDOT Division of Aviation* – www.ncdot.gov/aviation/uas. Morrisville, NC: NORTH CAROLINA DEPARTMENT OF TRANSPORTATION (accessed December 7, 2022).

Science Education Resource Center at Carleton College. *Teaching with Geopads*. https://serc.carleton.edu/research_education/geopad/index.html (accessed July 2022).

EXERCISES

1. Define the following terms and distinguish the differences between them.
    a. Unmanned aerial vehicle (UAV)
    b. Unmanned aerial system (UAS)
    c. Drone

2. What are the steps in a UAV/UAS/drone survey workflow?
3. What is the purpose of a ground control point in a UAV mapping project?
4. Explain what is meant by *georeferencing* and why GPS/GNSS is the preferred method for georeferencing in a UAS survey/mapping project.
5. What is Ground Sampling Distance?
6. Which of the following statements is correct regarding UAS surveys?
    a. A licensed operator is permitted to fly the UAS aircraft for which they are licensed, at any project location without any restrictions whatsoever.
    b. In case there are multiple flights to be conducted, it is not necessary to repeat checklist steps to prepare the aircraft for launch again.
    c. It is not necessary to recheck weather conditions before a flight launch if the weather has stayed the same for a week and the forecast last night showed no changes expected for the next few days.
    d. Physical ground control points can be constructed in many ways and there are no restrictions or rules as long as they are big and clear enough to be detected by a UAV camera lens during mission flight.
7. Discuss why maintaining a visual line of sight is important during a UAS mission.

10 Engineering and Mining Surveys

10.1 INTRODUCTION

Engineering Surveying, a geomatics discipline, provides the knowledge and skills essential for carrying out accurate geospatial measurements which are mainly used in construction projects. It is also commonly used in deformation monitoring in areas such as monitoring the performance and health conditions of completed physical infrastructure projects and related facilities. Project planning and design, and quality control and quality assurance are strictly observed to ensure project implementation according to design standards and tolerances.

In common practice, engineering surveys are applicable in completed land-based engineering infrastructure and construction projects such as drainage channels, tunnels, roads, bridges, tall buildings, railways, airports, and dams, among other human-made structures.

Engineering surveys are typically more stringent in accuracy requirements with tolerances for most linear measurements being a few millimeters, unlike in areas such as cadastral, boundary, and topographic surveys where tolerances of centimeters are allowed. Due to the ongoing evolution in industry dynamics, it is common to come across other granular subdivisions of Geomatics which still apply the fundamental principles of Engineering Surveying. Examples of such subdivisions include construction surveys, mining surveys, control surveys (both horizontal and vertical), topographic surveys, detail surveys, route surveys, building surveys, hydrographic surveys, and so on. Mining surveys, as part of applied precision science and engineering, advance the utility of geospatial metrics to guide mining activities while making use of the principles of surveying, geodesy, mining, and geology.

10.2 GENERAL CONSIDERATIONS

10.2.1 Fundamental Principles and Concepts

Engineering Surveying involves project planning and design, and the quality control and quality assurance to ensure project implementation according to design standards and tolerances. In deformation monitoring, it is used in monitoring the performance and health conditions of completed works of physical infrastructure and related facilities. Thus, engineering surveys are critical to the operation and maintenance of engineering structures.

Due to the short distances involved in practice, the principles of plane surveying apply to both engineering and mining surveys, treating the Earth as flat. In some industry settings, very highly accurate surveys are necessary to meet the demanding standards of precision, an aspect of industrial metrology. This is a clear departure from geodetic surveying, which involves extensive areas for which the curvature of the Earth must be considered.

Horizontal control through traversing and vertical control through leveling techniques are essential to carrying out engineering surveys. Chainage is used to give reference to a specific location from a given starting point when carrying out horizontal control. Horizontal alignment for infrastructure such as roads, railways, or tunnels requires horizontal control to achieve the design standards, guided by a network of known (reference) points. Vertical alignment is also necessary for engineering and mining projects because slopes or gradients must be set out to allow for accurate elevation differences, motion, and material flows within designed safety margins. Benchmarks and/or Temporary Benchmarks (TBMs), which are points of known elevation, are set up to aid in vertical control.

During preliminary planning stages and initial excavations, the tolerances may be in the order of several centimeters, but they are improved to millimeters during final works, for example, when setting out invert levels for concrete lining in a drainage channel or hydropower tunnel. It is standard practice to determine the most suitable instrument and the accuracy that any given engineering or mining surveying exercise demands. The common instruments used for horizontal control are optical solutions (total stations and electronic theodolites), mobile devices with GNSS (such as handheld GNSS or even smartphones), geodetic solutions (geodetic GNSS receivers), and laser precision instruments. For vertical control, automatic levels are the engineer's choice due to their ease of operation and high accuracy in determining orthometric heights (H), which are heights referred to the geoid as opposed to the less accurate ellipsoidal heights (h) obtained from satellite-based positioning using GNSS receivers.

Surveyors use checks to ascertain the accuracy of horizontal and vertical control. Having at least three known points to start with is recommended to ensure the redundancy required for a check. The checks can take the form of a mathematical formula, such as confirming if the sum of the interior angles measured for a polygon adds up to the expected value, or if a loop traverse returns a computed value that matches the known value at the starting point. If the error is small or relatively small, it is usually distributed proportionately among the occupied points, assuming a linear relationship. If the errors are gross and the instrument has been confirmed to be accurate through standard tests or calibration, then the exercise needs to be repeated.

10.2.2 Common Instruments and Accessories

The instruments and accessories shown in Figure 10.1 are commonly used in engineering surveys. Curves and slopes are normally set out precisely using optical survey instruments.

Being precise and easy to use, automatic levels have become the industry standard for vertical control. Experience is key to setting up the instrument on a tripod (Figure 10.2) and taking accurate readings on a vertically held leveling staff.

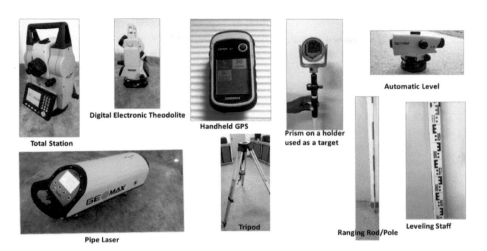

FIGURE 10.1 Common instruments and accessories for engineering surveys.

FIGURE 10.2 Students learning to operate an automatic level. (Photo credit: Nashon Adero, Taita Taveta University, Kenya.)

Determining elevation differences and deducing the reduced levels of the points of interest are achieved using two standard methods: Height of Collimation Method and Rise and Fall Method. The former is preferred as the faster method.

Digital levels, which are more expensive and sophisticated, use barcodes to give digital records of elevation, a highly convenient feature for engineering surveys when higher speed, accuracy, and precision are desired in a project.

A gyrotheodolite, or surveying gyro, is particularly very useful in mining surveys. It comprises a gyroscope mounted to a theodolite and is the main instrument for orientation in mine surveying and tunnel engineering, where there is no clear view of the sky and hence GNSS cannot work. A gyrotheodolite is used to determine the orientation of true north, except at or near (within 15° of) the poles, where meridians converge, and the east–west component of the Earth's rotation is not to the extent that can help obtain reliable results.

Lasers are the main precision instruments used for alignment in tunnels. They have the advantage of generating a sharp, precise, visible line of reference under the poorly lit underground environment characterizing tunneling and underground mining activities. Against this line of reference, surveyors can direct the vertical and horizontal alignments for excavations and estimates of quantities of "underbreaks" and "overbreaks."

10.2.3 Exposition Using a Case Study on Tunnel Surveys

A case study on tunneling has been selected to exemplify the rigor and granularity of engineering and mining surveys. Long tunnels may require teams to work from both ends and meet up mid-way, hence the stringent accuracy requirements to avoid gross errors or minimize the systematic errors that can result in the teams veering off horizontally and/or vertically—never to meet up at all with massive losses of investment. Surveyors must consider the differences in coordinate and reference systems for transboundary tunnels to ensure that mathematical calculations and error-minimization techniques are harmonized.

Tunnels may serve the needs of mining, railway transport, road transport, non-motorized transport, water transport, and hydropower production, among others. Notable tunnels and amazing products of engineering surveys in the world include, for example:

- The Delaware Aqueduct in the New York City water supply system (built 1939–1945)—the world's longest tunnel at 13.5 ft (4.1 m) wide and 85 miles (137 km) long.
- Gotthard Base Tunnel from Zurich to Milan—the world's longest rail tunnel (57.09 km) as of June 2016 (built from 1996), at the cost of US$10.3 billion.
- Seikan Tunnel connecting the Honshu and Hokkaido islands in Japan (53.9 km, 23.3 km under seabed), completed in 1988, at the cost of US$3.6 billion.
- Channel Tunnel (50.5, 37.9 km under the sea), constructed from 1987 to 1994 at the cost of £4.65 billion.

Due to their execution in restricted underground environments with neither natural lighting nor physical landmarks for approximate positioning, tunnel surveys qualify as one of the strictest types of engineering surveys. Air must be pumped along long tunnels for the safety of workers, who should be in full safety gear. Underground water is a common challenge in tunneling, hence the use of electric power to pump water out of the tunnels to facilitate survey procedures such as fixing reference points along the designed centerline and TBMs, usually using concrete nails. Measures must be taken to avoid cases of electrocution due to faulty wires that may be exposed to the water. Blasting of rocks for easier tunnel excavation is common practice. For safety, geological investigations are required to determine rock quality and the right reinforcement needed, which may vary from shotcrete to iron bars or iron plates.

Figure 10.3 shows an example of how laser beams are used to guide the excavation of a straight section of a tunnel with an excavation radius of 2.2 m. This example is drawn from a practical application case of constructing a headrace tunnel for hydropower production in Kenya, the Sondu-Miriu Hydropower Project, for which the final radius after lining with concrete was designed to be 2.1 m. More radial allowance is necessary during excavation to give room for sufficient thickness of concrete lining and for the movement of people, vehicles, machinery, and tools such as formwork of shutters and reinforcing bars.

Using traversing and leveling, targets are surveyed to be positioned at the correct (X, Y) locations and elevations (H) such that the laser beams are parallel and also tracing the correct slope according to the designed downstream/upstream slope of the tunnel section, 1 in 1,000 in this practical case. The second pair of targets coinciding with the chainage of H' provides a necessary check. Measurements from the reference laser beam, both diametrically and vertically, aid in setting out the measurements needed not only for accurate excavation and construction to replicate the designed properties, but also for volumetric estimates of earthworks. It can be appreciated from this example that both horizontal control and vertical control are critical and must be accurate to achieve such an amazing engineering feat.

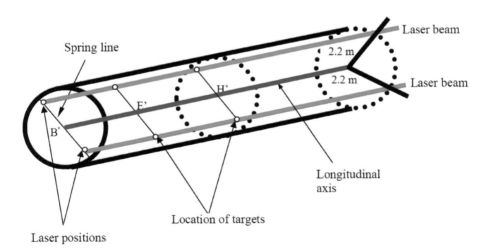

FIGURE 10.3   Straight section of Sondu-Miriu Hydropower Tunnel in Kenya. Drawn on site from surveying experience during civil works. (Credit: Nashon Adero.)

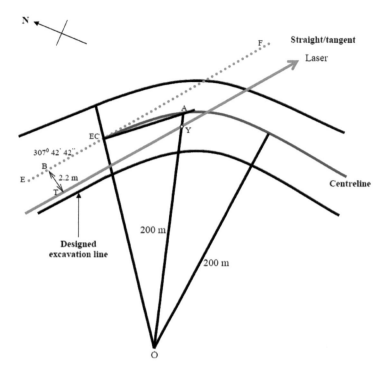

FIGURE 10.4 Curved section of Sondu-Miriu Hydropower Tunnel in Kenya. Drawn on site from surveying experience during civil works near the intake weir. (Credit: Nashon Adero.)

Figure 10.4 shows how in the same tunneling project a laser beam was used to set out the curved section of the tunnel. The exercise becomes more demanding because the designed curve geometry must be taken into account this time. As shown, the curve radius is 200 m, using the centerline as the reference as per the standard practice of setting out curves. The surveyor's challenge is to determine the offset AY, which gives the measurement from the laser beam to the point on the designed centerline. The surveyor needs to know the designed bearing of the straight/tangent to the curve at the point EC (known coordinates and chainage) and the chainage of A. As will be shown in the section for exercises, the mathematical equations for a straight line in rectangular coordinates (plane surveying), curve length, and chord length are routinely applied to arrive at the solution for the offset AY.

These underground application examples suffice for underground mining surveys as well because tunneling is a necessary undertaking in such projects.

10.3 TRANSFERRING CONTROLS

As a general rule, the surveyor should test the instruments to be used to ensure their accuracy, based on points of known (X, Y) coordinates and elevations (H). Optical solutions, GNSS, and lasers are commonly applied to transfer survey controls or reference points from one location to another. This step is necessary so that survey projects can be extended and executed within a controlled and location-specific framework of (X, Y) coordinates and elevations (H or Z). The tolerances vary, from relaxed decimeter or centimeter levels at the preliminary stage to stringent millimeter levels at the final construction stages. The process of increasing the density of control points or reference points in an area is also known as *densification*.

As a satellite-based solution, GNSS has made it easier to create (X, Y) control points without running long traverses from primary control points in a country, which used to be the case before the invention and commercial availability of GNSS services. Static positioning techniques are commonly applied to densify control points for construction. Differential real-time kinematic GNSS and software-based post-processing methods are also common, depending on the surveyor's needs-based choice and availability of GNSS continuously operating reference station (CORS) networks infrastructure.

Augmented GNSS services such as SBAS have helped to enhance the accuracy of GNSS positioning. CORS are part of the solutions that have been advanced to increase the accuracy of GNSS positioning in many countries. There is a promising future of GNSS and its prospects for application in obstructed and confined environments such as urban canyons, because of inventions such as multi-constellation satellite service capabilities and the renowned Z-Blade GNSS-centric technology.

Shafting and *plumblines* are a common means of vertical alignment for tunneling works. Benches also assist surveyors in transferring controls from the surface to underground, using a series of instrument stations. When using benches, traversing and leveling are the key procedures surveyors undertake to transfer horizontal and vertical control, respectively. Pegs are used for demarcating formation levels, marking them off with conspicuous spray paint. Other tools such as tape measure and strings are also useful and mostly applicable at the preliminary stages, for example, when guiding the course of excavation where a road should pass through.

10.4 CONSIDERATIONS FOR SETTING OUT AND DEFORMATION MONITORING

Setting out of engineering structures relies on a network of control points and offsets calculated in reference to specific points on reference lines. Offsets are linear distance measurements perpendicular to the reference line at a point. The designed centerline of a structure such as a road or a tunnel is used as the reference line by default. Vertical control is also necessary to set out the right elevations and slopes. Railways, for example, demand a very gentle slope, normally less than a degree, because of the low friction between the steel wheel and the steel rail. Roads can manage a higher gradient or grade. A surveyor must take such factors into account during setting out in a project.

On design documents, slope may be expressed as 1 in X, 1/X, or in degrees (Ø). In any case, slope = tan (Ø) = dy/dx (vertical difference divided by horizontal equivalent), which is 1/X in this example, also equal to the tangent of the slope angle. The slope can also be expressed as a percentage, which is the same as 100 tan (Ø).

Deformation monitoring involves carrying out systematic, precise, and periodic measurements on a structure in order to detect (usually small and gradual) displacements as a function of time in the structure away from the originally established positions, shape, or geometrical relationships—normally due to induced stresses. Deformation monitoring is key to preventive maintenance. With technological advances, such as Big Data, Machine Learning, and Artificial Intelligence, predictive maintenance is becoming a reality as well.

Examples of structures to be monitored include bridges, dams, buildings, tunnels, road and rail infrastructure, fuel storage tanks, and pipelines. Precise survey measurements help detect lateral and/or vertical shifts/movements/subsidence. The measurement techniques employed can be as simple as measurements of radials in a tunnel using a leveling staff, such as within an excavated tunnel, to complex ones yielding sub-centimeter (millimeter) measurements using precision instruments and photogrammetry techniques. The survey methods applied to monitor structural deformations make use of traversing, leveling, laser surveys, photogrammetry/drone and LIDAR-assisted generation of point clouds, or satellite-based positioning methods (GNSS/Global Positioning System [GPS]/satellite images).

In summary, a surveyor needs to consider the following instrumentation and procedures for a complete deformation monitoring schedule:

- *Instrumentation:* Optical instruments, GPS/GNSS, and imagery/photogrammetric set ups.
- Reference points are accurately determined.
- Monitoring points are placed on the structure with receptors, for example, calibrated reflectors.
- Monumenting of the control points as a cautionary measure against the displacement of reference points.
- Taking periodic measurements from the control points, used as reference stations, and taking readings to the monitoring points on the structure to obtain data on any changing geometries.
- Deformation analysis based on well-grounded theory to reflect on the ideal condition, against the observed conditions at different epochs/times.
- Reporting of the results and recommendations for necessary preventive interventions.

10.5   MINING SURVEYS

10.5.1   THE FUTURE OF MINING SURVEYS

Minerals are driving the future of industrialization and human civilization. Battery and fuel technologies, for example, are gaining policy relevance in a world experiencing a growing uptake of sophisticated consumer electronics and electric vehicles. Environmental responsibility in the mining sector is a key global issue in the pursuit of the UN Sustainable Development Goals (especially Goal 7, Goal 13, and Goal 15). Decarbonization and meeting net-zero emission targets in the face of climate change are compelling policy goals.

Emboldened by recent global developments, mining scholars have rightfully argued that, in terms of importance to society, mining should be accorded a weight that is no less than the weight of farming, fisheries, and forestry. Mining surveys, as a result, are gaining prominence. Issues of labor aside, automation is gaining importance in the global mining industry because it is key to safety enhancement. Automation of operations in the mining sector including machine guidance requires accurate geospatial positioning techniques. Research on the bleeding edge also promises deep-sea mining as the next frontier of mining, making mining surveys even more critical in future.

10.5.2 CLASSICAL AND EMERGING FEATURES OF MINING SURVEYS

Mining surveys involve applied precision science and engineering that generates the geospatial metrics needed to guide mineral exploration and mining activities while utilizing the principles of surveying, geodesy, mining, and geology. Mining surveys can be considered as a special subclass of engineering surveys applied in mining environments. The life cycle of any mining enterprise, from exploration to post-closure activities, presents many phases in which the role of surveying and geospatial techniques and technologies remains critical. Spatial metrics are key to policy decision support, safety monitoring, mineral resource exploration, mine planning, mine design, determining land-related mining rights with exactitude, quantitative estimates of earthworks or materials on site, and post-mining rehabilitation or landscape restoration.

Deposit processing involves the exploration, evaluation, and determination of the reserves of a deposit and supports a mining plan in regard to optimal use of the reserves. Approval procedures are a precondition when exploring and extracting mineral resources. They are the basis for the acquisition of a mining permit and access to the land where mining takes place.

Traversing and leveling are classical methods that continue to find application in mining fields. Theodolites, total stations, and levels are commonly used, complemented by laser precision instruments. Figure 10.5 shows a typical underground activity that requires the application of surveying techniques. Horizontal and vertical alignment of the reinforcing iron bars inside the tunnel were accomplished using a total station and a level. An adjustable 15-m-long formwork made of steel was aligned first to assume the designed shape and orientation of the tunnel, after which the iron bars were laid on it to take on the shape and orientation of the formwork before the formwork was collapsed and shifted to a new position. The aim was to end up with a tunnel of a circular shape once lined with reinforced concrete.

A *gyrotheodolite* is used for orientation in underground environments. Laser surveys (including airborne laser scanning) and the common optical solutions in engineering surveys are applied to deliver mining-related solutions as well. Slope planning, pegging, setting out, deformation monitoring, and determining rights and legal liability by mining blocks are key examples. Because land and mining are strongly interconnected, cadastral surveys are critical to determining mining rights on land. Countries are advancing toward digital 3D cadaster so that mining rights

FIGURE 10.5 Reinforced concrete lining of a tunnel section using optical solutions—a total station and a level. (Photo credit: Nashon Adero.)

can be better managed, including the effects of underground mining on the structures owned by neighboring communities.

A *borehole camera* is part of the highly useful equipment in mining surveys. The 3D spatial coordinates of borehole data can be analyzed using Geographic Information System (GIS) techniques and artificial neural networks to reveal characteristics that are key to detecting geohazards.

A borehole camera has the following key functions:

• Ascertaining groundwater conditions—shows location, color, consistency, and amount of precipitates
• Viewing subsurface conditions
• Fracture logging
• Documenting cracks or holes and leaking joints
• Observing hole offsets and blockages
• Viewing and recording the stratigraphy and lithology of the accessed units.

Airborne geophysical surveys produce geophysical maps and are commonly conducted to map out the mineral potential of a territory. Control points are established to ensure that the final map can be georeferenced, hence enriching the final products with actionable location-based intelligence. On mining sites are key structures such as storage tanks and dams, which do need regular deformation monitoring to ensure safety. Blasting and other highly vibratory activities on mining sites due to heavy machinery make subsidence monitoring a crucial undertaking.

Remote sensing and GIS are together potent tools for assessing, quantifying, and monitoring the effects of mining on land and environment. They are important sources of the data and information that is increasingly required for sound decision support and policy development for sustainable mining practices in the broader mining–environment–society nexus. Advances in the resolution of satellite imagery are delivering more resourceful data for land use and land cover assessments across mining areas. Initiatives such as Digital Earth Africa and Digital Earth Australia are providing analysis-ready data and decision-ready data from processed satellite imagery, making it even easier to generate spatial models that support decisions and policies on sustainable mining and reclamation of closed mining sites. Sentinel (10 m), PlanetScope (3 m), Pleiades (0.5 m), WorldView-3 (0.3 m), and hyperspectral images of sub-meter spatial resolutions exemplify the innovations pushing the boundaries of satellite-based imagery solutions for the mining sector.

Land subsidence is a key risk that should be monitored keenly on mining sites. *Differential radar interferometry* or *Differential Interferometric Synthetic Aperture Radar* (DInSAR) has been a powerful means of monitoring small changes that lead to land subsidence in mining areas. This method, based on active remote sensing, is more effective and economical over large areas than the less effective point-by-point measurements obtained using classical optical solutions. These technological advances are a great boost to subsidence engineering.

Augmented GNSS delivers centimeter-level accuracy for automation and machine guidance in mining. This solution is also applicable to route planning and fleet management, which are key to enhancing productivity in busy mining operations. With offshore prospects in deep-sea exploration for minerals, surveys of the sea-floor morphology or bathymetry make use of multibeam acoustic and airborne laser systems, nautical charts, and echo sounders for depth measurements. GNSS still finds application in navigation for safety and for transport of cargo during such exploration missions.

Technological innovation is enhancing offshore mineral exploration prospects through autonomous underwater vehicles. For sea-floor exploration, navigation and sampling use *autonomous robotic systems* that can work under high pressure, low temperature, and total darkness. Intelligent response to sound frequencies that trigger the dropping of weights used during immersion enables the return of the vehicles to the surface, relying on their buoyancy. Several case studies of these efforts exist with the GEOMAR Helmholtz Centre for Ocean Research Kiel, Germany, acting as a key example in deep-sea exploration for minerals.

Software developments continue to boost model development for planning and design related to mining. Block models can nowadays be readily generated using specialized commercial software such as Surpac and MineSight 3D. Automated cartographic solutions for mapping and planning at scale add to the long array of benefits of software development.

10.5.3   Policy, Legal and Regulatory Aspects in Mining Surveys

Mining rights are tightly linked to sensitive issues and questions around land tenure and the physical environment, hence to land and environmental rights as well. Many studies across the world have established extensive violations of human rights in the mining sector, a key example being the case of Taita Taveta, Kenya, which was confirmed in 2016 through a public inquiry led by the Kenya National Commission on Human Rights. The inquiry found that the violations were widespread, from land rights, environmental rights, gender rights, children rights, to labor rights. Key to gaining a shared understanding of the extent of such violations of rights and resolving them is the geospatial information from accurate mining surveys and mapping.

Conducive policies, laws, and regulations are needed to resolve and manage conflicts in the mining sector. Mining surveys provide the spatially explicit metrics and parameters required to accurately allocate, enforce, and monitor land-related mining rights. Maps generated using various techniques (e.g., aerial photogrammetry, laser scanning, satellite imagery, topographical surveys, cadastral surveys, and GIS) with data from various sources complement the datasets and information needed for sound decision support when administering such crucial matters of land and mining rights, all the way to minimizing the negative environmental effects of mining and ensuring adequate rehabilitation after mine closure.

The evolution of mining legislation has realized significant milestones, not least in Africa. In Kenya, for example, a modern mining law introduced in 2016 has been referred to as Africa's most progressive mining law. The Mining Act of 2016 specifies how mining rights should be administered based on a digital mining cadaster and units of blocks defined by graticules spaced 15″ apart. This Kenyan example confirms the important role of mining surveys and mapping in managing and administering mining activities and mining rights.

The illustration of a mining block in Figure 10.6 serves as a suitable example on how to translate legal and policy aspects into practice when issuing mining rights or

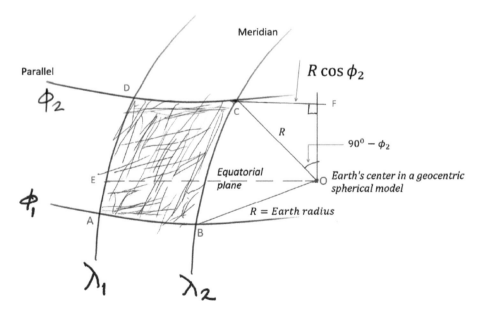

FIGURE 10.6 A pseudoquadrilateral ABCD representing a mining block based on a geocentric spherical model of the earth.

licenses attached to land parcels. ABCD is the pseudoquadrilateral representing a mining block. In the Kenya Mining Act of 2016, for example, such a block is bounded by meridians spaced 15 arc seconds apart and parallels spaced 15 arc seconds apart. As shown based on a spherical Earth model, R is the radius of the Earth, estimated to be 6,371 km.

Generally, the mining block in Figure 10.6 is interpreted as follows:

- Arc length for AB or CD = (longitude difference, i.e., $\lambda_2 - \lambda_1$ in degrees/360) × 2(RcosΦ), where Φ is the latitude (in degrees) of the circle of latitude, along which the longitude difference is measured. Near the equator (OE), RcosΦ is almost equal to unity.
- Arc length for AD or BC = (latitude difference, i.e., $\varphi_2 - \varphi_1$ in degrees/360) × 2πR, because all meridians are great circles such that R is the Earth radius always.

BIBLIOGRAPHY

Adero, N.J. 2003. The role of surveying in tunnelling: Practical experience. Technical Report documenting professional experience as a Tunnel Surveyor, Sondu-Miriu Hydropower Project, Kenya, 2001–2003.

Drebenstedt, C. and R. Singhal (eds.). 2014. *Mine Planning and Equipment Selection.* Switzerland: Springer Int. Publishing.

GoK - Government of Kenya. 2016. *The Mining Act (No. 12 of 2016). Kenya Gazette Supplement No. 71 (Acts No. 12).* Nairobi: Government Printer.

Hong-lan, W. 1987. Analysis of the motion of a gyro-theodolite. *Applied Mathematics and Mechanics,* 8: 889–900. doi:10.1007/BF02019527.

KNCHR – Kenya National Commission on Human Rights. 2017. *Public Inquiry Report on Mining and Impact on Human Rights: Taita Taveta County, 2016.* Nairobi: Top Quality Converters Co.

Schofield, W. and M. Breach. 2007. *Engineering surveying, Butterworth-Heinemann,* pp. 519–533. ISBN 0-7506-6949-7.

Spectra Geospatial. 2022. *Z-Blade. Product Details.* https://spectrageospatial.com/z-blade.

EXERCISES

1. A straight segment of a railway has a design gradient of 0.5°. You are required to survey from the downstream end at chainage 295 + 125 to the upstream end at chainage 295 + 140. The upstream level should have a reduced level of 597.000 m.
   a. Find X if the plan drawing expresses the railway gradient as 1 in X.
   b. Calculate the correct staff reading at the downstream end if a backsight reading of 3.025 m has been taken on a TBM whose reduced level is 594.945 m.
   c. List the survey instruments, tools, and accessories that would be suitable for this exercise, grouping them by functions under horizontal control and vertical control.

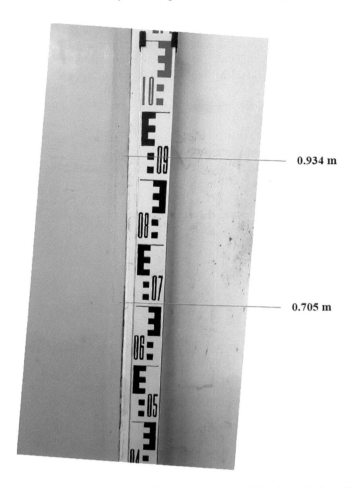

0.934 m

0.705 m

FIGURE 10.7 Note that a leveling staff is used together with a level. It is held vertically on the target point whose height is being measured. There are major graduations of 100 mm, minor graduations of 10 mm forming small blocks, and some of the 50 mm blocks out of the major 100 mm block are joined to form (inverted) E patterns.

2. Make a simplified illustration on how to take readings off a leveling staff. A typical solution is shown in Figure 10.7.
3. A surveyor uses a total station whose accuracy is specified in the operator's manual as 10 mm + 10 PPM. The instrument measures a chord length to be 150.030 m where the intersection angle is 600 for a curved tunnel of radius, R = 150 m.

   Based on the information above:
   a. Determine if this total station is fit for use on the site.
   b. Advise the engineers on the action to take based on this finding.
4. Refer to the diagram in Figure 10.8 showing the geometry of one curved upstream section of a water tunnel. It is based on the Sondu-Miriu

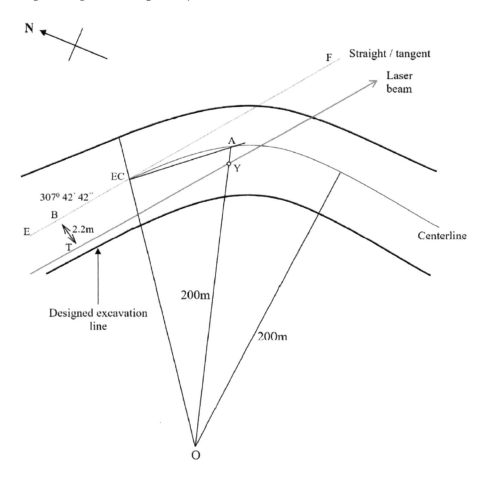

FIGURE 10.8   Curved section of a water tunnel.

Hydropower Project in Kenya. In the curved section of $R = 200\,\text{m}$, the engineering surveyors used formwork spans measuring only $3\,\text{m}$ in length, a choice that can be mathematically proven given $R = 200\,\text{m}$, because the curve length and chord length for such a short span do not vary significantly, but over several spans the errors can accumulate significantly if not checked. Solve for the offset distance AY to the designed centerline. (Hint: To get accurate results, the intermediate answers obtained should not be rounded off).

5. Zoom in to Taita Taveta University, Kenya, from any digital map service like Google Earth. Sketch a mining block around it bound by 15 arc seconds, as provided in the Kenya Mining Act of 2016. Prove that the linear dimensions of the sides of the block measure about $460\,\text{m}$. Based on this, deduce the number of mining blocks obtainable from an area of $800\,\text{ha}$.

Part IV

Proposal Development

11 Estimating Project Costs

11.1 INTRODUCTION

In order to estimate project costs, it is necessary to visualize procedures that must be accomplished. There must be a concept, mental or written, as to what is required to complete the project. For most projects, the starting point is the scope of work—a statement outlining general job specifications as provided by the client. This serves as the basis for project design. In this chapter, typical examples are provided in Sections 11.4.2, 11.5.2, and 11.6.2.

With a logical project plan in mind, it is possible to estimate man-hour and material needs and apply cost factors. Because hourly labor rates, equipment rental rates, overhead, and profit margin vary widely, it is necessary to estimate costs based on a specific scheme. Different methods are used to procure surveying and mapping services, depending on whether the client is a government or private entity. For simplicity in this chapter we will assume a general case of contracting with a government entity. This will provide a common reference for examining the topic of cost estimation with the goal of underscoring the principles of lump-sum cost estimation on the basis of unit prices (or rates). Exactly what are those unit prices, and how are they formulated?

In the U.S. federal government, professional architectural, engineering, planning, and related surveying services must be procured under the Brooks Architect-Engineer Act, Public Law 92–582 (10 US Code 541–544). The Brooks Act requires the public announcement of requirements for surveying services, and selection of the most highly qualified firms based on demonstrated competence and professional qualifications. After selection, negotiation of fair and reasonable contract rates for the work is conducted with the highest qualified firm. Procedures for obtaining services are based on a variety of federal and departmental regulations. For instance, two types of contracts that are used for surveying services include firm-fixed-price (FFP) contracts and indefinite delivery contracts (IDCs). FFP contracts are used for moderate to large mapping projects (e.g., greater than $1 million) where the scope of work is known prior to advertisement and can be accurately defined during negotiations. However, fixed-scope FFP contracts are rarely used because most government mapping work involving surveying services cannot be accurately defined in advance.

IDCs (once termed "open-end" contracts) have only a general scope of work, for example, "GNSS surveying services in Southeastern United States." When work arises during the term of the contract, task orders are written for performing that specific work. Task orders are then negotiated using a unit rate "Schedule" developed for the main contract. Negotiations are focused on level of effort and duration of the project. The scope is sent to the contracting firm who responds with a time and cost estimate, from which negotiations are initiated.

11.2 ELEMENTS OF COSTING

The main elements of (or factors for) cost estimates for surveying services are as follows:

   a. *Labor:* One of the most significant production factors in a surveying or mapping project relates directly to hours expended by highly qualified professionals and technicians. Amount of work that personnel will conduct is characterized as direct labor. It is convenient to express work in hours because it provides a per unit cost basis for estimating purposes. Cost per hour of personnel can be obtained from regional wage rates or from negotiated information supplied by the contracting firm. These can be applied to the estimated work hours to arrive at a project cost.
   b. *Capital equipment:* Another significant factor relates to the capital equipment that technicians operate during the work hours. Depending on project type, such equipment could include aircraft, GNSS receivers, total stations, CADD workstations, plotters, scanners, computers, film processors, and so forth. The equipment costs can be arrived at through hourly rental during project phases or through acquisition.
   c. *Deliverables:* A list of project delivery items (end products) should be supplied by the client. This is necessary to ensure an accurate cost estimate, for example, on the basis of the required materials, supplies, labor, and miscellaneous. Products should be specified in the contract, and the number of copies or sets to be furnished must be stated.

A number of methods are used for estimating and scheduling surveying services in a fixed-price or IDC contract. One of the methods is a "daily rate" basis, although hourly rates for personnel labor may be preferred by some entities. A daily rate basis is the cost for personnel or equipment over a nominal 8-hour day. In some cases, a composite daily rate may be estimated and negotiated for a full field crew (including all personnel, instrumentation, transport, travel, and overhead). The crew personnel size, total stations, number of GNSS receivers deployed, vehicles, and so forth must be explicitly indicated in the contract specifications, with differences resolved during negotiations.

Factors for estimating costs can be itemized using the analysis shown in Table 11.1 (adapted from the U.S. Army Corps of Engineers [USACE] Manual Number EM 1110–1–1005, 13–4, Jan. 2007).

11.3 UNIT PRICE SCHEDULES

The various personnel and equipment cost items like those shown in the previous section are used as a basis for negotiating fees for individual line items in the basic IDC contract. During negotiations, individual components of the price proposal may be compared and discussed. Differences are resolved in order to arrive at a fair and reasonable price for each line item. The contract may schedule unit prices based on variable crew sizes and equipment. Examples of negotiated unit price schedules are

TABLE 11.1
Sample Cost Estimating Analysis

Item	Description
I	Direct labor or salary costs of survey technicians: includes applicable overtime or other differentials necessitated by the observing schedule
II	Overhead on direct labor[a]
III	General and administrative (G&A) overhead costs (on direct labor)[a]
IV	Direct material and supply costs
V	Travel and transportation costs: crew travel, per diem, airfare, mileage, tolls, and so forth. Includes all associated costs of vehicles used to transport personnel and equipment
VI	Other direct costs (not included in G&A): includes survey equipment and instrumentation, such as total stations and GNSS receivers. Instrument costs should be amortized down to a daily rate, based on average utilization rates, expected life, and so forth. Some of these costs may have been included under G&A. Exclude all instrumentation and plant costs covered under G&A, such as interest
VII	Profit on all of the above (computed/negotiated)

[a] These may be combined into a single overhead rate.

shown in Tables 11.2 and 11.3. The contract specifications would contain the personnel and equipment requirements for each line item. Here are some of the line items explained.

a. *Personnel and crew line items:* Individual line items are explicitly defined in the IDC contract specifications. For example, the specifications must define what instrumentation, if any, is included in a "2-Person Survey Party" item in Table 11.3.

b. *Overtime rates:* Overtime rates may be applied during emergency operations. They may be estimated based on nominal 40-hour weeks (8-hour or 10-hour workdays). The $28.86 overtime rate for the "Party Chief" (Table 11.3) is based on 1.5 times a base hourly rate of $19.24. The daily rate of $384.80 is determined from $19.24/hour × 8 hour/day × 150% overhead rate.

c. *Mob/demob:* Mob/demob times would be applied to the time estimates for personnel and equipment. The mob/demob rates would depend on the job location (there might be cases where the work site is the same for the entire contract period, then a fixed rate would be applicable).

d. *Miscellaneous items:* Generally, it might be preferred to lump miscellaneous supplies into a crew rate or to include it in overhead. For instance, the $10.00 line item in Table 11.3 for "Materials" could have been included in the contract overhead. If there is a major requirement for supplies on a task order, then this can be negotiated accordingly, for example, "200 monuments with bronze discs."

TABLE 11.2
Sample Unit Price Schedule for GNSS Services

Line Item	Unit	Rate
Registered/Licensed land surveyor (office)	Daily	$497.31
Registered/Licensed land surveyor (field)	Daily	$459.22
Professional geodesist computer (office)	Daily	$415.76
Engineering technician (CADD draftsman) (office)	Daily	$296.00
Civil engineering technician (field supervisor)	Daily	$245.00
Supervisory GNSS survey technician (field)	Daily	$452.73
Surveying technician (GNSS instrument man/recorder)	Daily	$374.19
Surveying aid–rodman/chainman	Daily	$246.94
One-person GNSS survey crew (two receivers–one vehicle–travel)	Daily	$1,043.35
Two-person GNSS survey crew (two receivers–one vehicle–travel)	Daily	$1,415.76
Three-person GNSS survey crew (three receivers–two vehicles–travel)	Daily	$1,868.05
Four-person GNSS survey crew (four receivers–three vehicles–travel)	Daily	$2,234.72
Additional GNSS receiver	Daily	$100.00
Additional survey vehicle	Daily	$40.00
Station monuments—standard concrete monument	Each	$25.00
Station monuments—deep rod vertical monument	Each	$950.00
Bluebooking	—	$500.00

Source: Adapted from EM 1110–1–1003 of the U.S. Army Corps of Engineers, 12–4, July 2003.
Note: Scheduled prices include overhead and profit (these could be listed separately if desired). GNSS survey crew includes all field equipment; auxiliary data loggers; tripods; and computers needed to observe, reduce, and adjust baselines in the field. Per diem is included. (The contract scope of work will specify items that are included with a crew, including GNSS receiver quality standards.)

Cost estimates are required for each line item in the unit price schedules (Tables 11.2 and 11.3). The estimates must be sufficiently detailed such that a "fair and reasonable" price is reached in a contract. The cost computation examples in the following sections are representative of the procedures that can be applied in preparing a fair and reasonable price. The costs and overhead percentages are shown for illustration only—they are subject to considerable variations, depending on a project, geographic location, and contractor-dependent variations (e.g., audited general and administrative [G&A] rates could range from 50% to 200%).

11.4 GNSS SURVEY COST ESTIMATING

11.4.1 Sample Cost Estimate for GNSS Services

The example in Table 11.4 shows the cost computation for a three-man GNSS survey crew, a representative of the procedure used in preparing the contract price schedule

TABLE 11.3
Sample Unit Price Schedule for Topographic Surveying Services

Line Item	Unit	Rate
Supv prof civil engineer	Daily	$795.60
Supv prof land surveyor	Daily	$681.20
Registered land surveyor	Daily	$572.00
Civil engr tech	Daily	$364.00
Cartographic tech (includes CADD operator)	Daily	$332.80
Stereo plotter operator (includes photogrammetric softcopy operator)	Daily	$455.52
Engineering/Cartographic aid	Daily	$309.92
GIS system analyst	Daily	$582.40
GIS database manager	Daily	$542.88
GIS technician	Daily	$343.20
Party chief	Daily	$384.80
Party chief (overtime)	Hourly	$28.86
Instrument person	Daily	$291.20
Rodman–chainman–laborer	Daily	$234.00
4-person topographic survey party	Daily	$1,196.00
3-person topographic survey party	Daily	$904.80
2-person topographic survey party	Daily	$665.60
1-person topographic survey party	Daily	$502.98
Mob and demob of survey party	Per project	$988.00
Total station equipment cost—cost per instrument and data collector, per day	Daily	$50.00
GNSS equipment cost—cost per receiver, per day	Daily	$75.00
Field computing PCs and software	Daily	$50.00
Misc. Items		
Materials (PVC, steel fence posts, rebar, misc.)	Daily	$10.00
Mileage—4 wheel truck	Per mile	$0.60
Per diem (estimate actual costs on each order)	Daily	—
Profit (use 10.5% for all task orders)	—	—

Source: Adapted from EM 1110–1–1005 of the USACE, 13–5, Jan. 2007.

(Table 11.2). Larger crew/receiver size estimates would be performed similarly. Associated costs for GNSS receivers, such as insurance, maintenance contracts, interest, and so forth, are presumed to be indirectly factored into a firm's G&A overhead account. If not, then such costs must be directly added to the basic equipment depreciation rates. Other equally acceptable accounting methods for developing daily costs of the equipment may be used.

TABLE 11.4
Sample Computation for a Three-Man GNSS Survey Crew[a]

Labor

Supervisory survey tech (party chief)	$42,776.00/year	
Overhead on direct labor (36%)	$15,399.36/year	
G&A overhead (115%)	$49,192.40/year	
Total:	$107,367.76/year	$411.57/day
Survey technician—GNSS observer	$35,355.00/year	
@151% O/H (36% + 115%)	$88,741.05	$340.17/day
Survey aid	$23,332.00/year	
@151% O/H (36% + 115%)	$58,563.32	$224.49/day

Total labor cost/day:	$976.23
Travel (Nominal Rate)	
Per diem: 3 persons @ $88/day	
Total travel cost/day:	$264.00
Instrumentation & Equipment	
GNSS system: 3 geodetic quality receivers (static or kinematic), batteries, tripods, data collectors, and so forth.	
$40,000 ea or $120,000 @ 4 years @ 100 day/ year	$300.00/day
Total station: data collector, prisms, and so forth	
$32,000 @ 5 years @ 120 day/year (rental rate: $60/day)	$53.00/day
Survey vehicle: $40,000 ea @ 6 years @ 225 day/year plus O&M @ 2 reqd.	$80/day
Misc materials (field books, survey supplies, etc.)	$25/day
Total inst. and equipment cost/day:	$458.00
Subtotal:	$1,698.23
Profit @ 10%:	$169.82
Total estimated cost per day:	$1,868.05

Note: Similar computations are made for other line items in the price schedule.

[a] Three receivers, auxiliary equipment, two vehicles, laptops, and adjustment software.

If a cost-per-work unit fee structure is desired on an IDC, typical work unit measures on a GNSS contract might be cost per static point or cost per kinematic point. Cost per GNSS stations were commonly used during the early days of GNSS (mid-1980s) when GNSS receivers cost $150,000 and only 3–4 hours of satellite constellation was available each day. Today there is little justification for using work unit costs for pricing GNSS surveys.

11.4.2   SAMPLE SCOPE OF WORK FOR GNSS SERVICES

11.4.2.1   Specifications and Accuracy Standards

In the scope of work for services, specifications and standards for surveying work should make maximum reference to existing standards, publications, and other references. In the case of government contracts, the primary reference might be a manual. The U.S. government policy prescribes maximum use of industry standards and consensus standards established by private voluntary standards bodies, in lieu of government-developed standards. Consider, for example, this policy statement in a government manual: the EM 1110–1–2909:

Specifications for surveying and mapping shall use industry consensus standards established by national professional organizations, such as the American Society for Photogrammetry and Remote Sensing (ASPRS), the American Society of Civil Engineers (ASCE), the American Congress on Surveying and Mapping (ACSM), or the American Land Title Association (ALTA). Technical standards established by state boards of registration, especially on projects requiring licensed surveyors or mappers, shall be followed when legally applicable.

11.4.2.2   Sample Scope of Work

The following is an example scope of work for GNSS surveying services. Included in this example is the letter request for proposal to the IDC contract.

SAMPLE LETTER REQUEST FOR PROPOSAL

Engineering Division
Design Branch
Sea Systems, Inc.
3456 Northwest 27th Avenue
Pompano Beach, Florida 33069-1087
SUBJECT: Contract No. DACW17–98-D-0004
Gentlemen:
Enclosed are marked drawings depicting the scope of work requested for the following project:
One-Year Monitoring Beach Erosion Survey
Canaveral Harbor, Florida (Survey 99–267)

General Scope. Furnish all personnel, equipment, transportation, and materials necessary to perform and deliver the survey data below in accordance with the conditions set forth in Contract No. DACW17–98-D-0004. All work shall be accomplished in accordance with the manuals specified in the contract.

Your attention is directed to the Site Investigation and Conditions Affecting the Work clause of the contract. After we have reached agreement on a price and time for performance of the work, neither the negotiated price nor the time for performance will be exchanged as a consequence of conditions at the site except in accordance with the clause. Costs associated with the site investigation are considered overhead costs which are reimbursed in the overhead rates included in your contract. Additional reimbursement will not be made.

a. *Scope of work:* Hydrographic and topographic monitoring data shall be collected for CCAFS-29, CCAFS-30, CCAFS-33 through CCAFS-42, BC-5 through BC-14, and DEP R-0 through DEP R-18. The area is shown on enclosure 1, USGS quads. Enclosure 2 is the control monument descriptions and profile line azimuth. Enclosure 3 is the technical requirements for the surveys.

b. *Data processing:* The Contractor shall make the necessary computations to verify the accuracy of all measurements and apply the proper theory of location in accordance with the law or precedent and publish the results of the survey.

c. *CADD:* The survey data shall be translated or digitally captured into Autodesk Land Desktop according to the specifications furnished.

d. *Digital geospatial metadata*: Metadata are "data about data." They describe the content, identification, data quality, spatial data organization, spatial reference, entity and attribute information, distribution, metadata reference, and other characteristics of data. Each survey shall have metadata submitted with the final data submittal.

e. *Compliance:* Surveying and mapping shall be in strict compliance with the Minimum Technical Standards set by Florida Board of Professional Surveyors and Mappers.

The completion date for this assignment is 60 days after the Notice to Proceed is signed by the Contracting Officer.

Contact Design Branch at 907-223-1776 for assistance, questions, and requirements.

You are required to review these instructions and make an estimate in writing of the cost and number of days to complete the work. Please mark your estimate to the attention of Chief, Design Branch.

This is not an order to proceed with the work. Upon successful negotiation of this delivery order the Contracting Officer will issue the Notice to Proceed.

Sincerely,

Enclosures

(Source: Adapted from USACE EM 1110–1–1003, 12–8–12–9.)

SAMPLE SCOPE OF WORK

TECHNICAL QUALITY CONTROL REQUIREMENTS

ONE-YEAR MONITORING BEACH EROSION SURVEY

CANAVERAL HABOR, FLORIDA

(SURVEY 99–267)

1. *Location of work:* The project is located in Brevard County at Canaveral Harbor, Florida.

2. *Scope of work:*
   a. The services to be rendered include obtaining topographic and hydrographic survey data (x, y, z) and CADD data for 47 beach profile lines.
   b. The Contractor shall furnish all necessary materials, labor, supervision, equipment, and transportation necessary to execute and complete all work required by these specifications.

    c. Rights-of-Entry must be obtained verbally and recorded in the field book before entering on the private property. Enter the name and address of property owner contacted for rights-of-entry.

    d. Compliance. Surveying and mapping shall be in strict compliance with the Minimum Technical Standards set by Florida Board of Professional Surveyors and Mappers.

    e. Digital Geospatial Metadata. Metadata are "data about data." They describe the content, identification, data quality, spatial data organization, spatial reference, entity and attribute information, distribution, metadata reference, and other characteristics of data. Each survey shall have metadata submitted with the final data submittal.

    f. All digital data shall be submitted on CD ROMs.

    g. *Existing data:* The Contractor shall be furnished digital terrain model (DTM) files and existing sheet layout of previous monitoring survey. The Contractor shall utilize this information to perform survey comparisons.

3. *Field survey effort:*

    a. *Control:* The Horizontal datum shall be NAD 83 and the vertical datum shall be NAVD 88. All control surveys shall be Third-Order, class II accuracy.

    b. The basic control network shall be accomplished using precise differential carrier-phase GNSS and differential GNSS baseline vector observations.

    c. Network design, station and baseline occupation requirements, for static and kinematic surveys, satellite observation time per baseline, baseline redundancies, and connection requirements to existing networks, shall follow the criteria given in the manual.

    d. GNSS-derived elevation data shall be supplied in reference to the above said datum. Existing benchmark data and stations shall be used in tandem in a minimally constrained adjustment program to model the geoid. The GNSS plan shall be submitted and approved prior to commencing work.

    e. Establish or recover one horizontal and vertical control monument for each profile line. The established position for each monument recovered shall be utilized and new positions shall be established for any new monuments established. The GNSS network (if required) shall commence from the control shown on Enclosure 2. All established or recovered control shall be fully described and entered in a field book. All control surveys shall be Third-Order, class II accuracy. The Contractor shall submit the field data and abstracts for the control networks for computation before commencing the mapping.

    f. All horizontal and vertical control (double run forward and back) established shall be a closed traverse or level loop, with Third-Order accuracy. All horizontal and vertical control along with baseline layouts, sketches, and pertinent data shall be entered in field books.

    g. All monuments, survey markers, and so forth, recovered shall be noted on the copies of control descriptions. Control points established or recovered with no description or out-of-date (5 years old) description shall be described with sketches for future recovery use.

h. All original field notes shall be kept in standard pocket-size field books and shall become the property of the government. The first four pages of the field books shall be reserved for indexing and the binding outside edge shall be free of all marking.

i. *Beach profiles:* Recover or establish one (1) horizontal and vertical control monument for the beach profiles. Utilize the coordinates, elevations, and azimuths shown on Enclosure 2.

j. Obtain data points (X, Y, Z) on 10-foot ranges (land), all breaks in grade greater than 1 foot vertically, vegetation line, tops and toes of dunes, seawalls, or other man-made features along the profile line.

k. *Breakline:* Breaklines shall be located for all natural and man-made features as needed. The breaklines shall be located with X, Y, and Z and identified.

4. *Data processing:* The Contractor shall make the necessary computations to verify the correctness of all measurements and apply the property theory of location in accordance with the law or precedent and publish results of the survey. The Contractor shall submit advance copies of the horizontal control so that final coordinates can be computed before commencing mapping. Compute and tabulate the horizontal and vertical positions on all work performed. Review and edit all field data for discrepancies before plotting the final drawings.

a. Furnish X, Y, Z, and descriptor ASCII file for each profile line and one X, Y, Z, and descriptor ASCII file with all data included for each area.

5. *CADD:* The survey data shall be translated or digitally captured into Autodesk Land Desktop (LDT) according to the specifications furnished.

a. *Global origin:* The LDT 3-D design file shall be prepared with a global origin of 0, 0, 2147483.65.

b. *DTM data:* The Contractor shall develop and deliver a surface model of the area and the model file shall have the .dtm extension. The DTM shall be developed from the collected data. Breaklines should include ridges, drainage, road, edges, surface water boundaries, and other linear features implying a change in slope. The surface model shall be of adequate density and quality to produce a 1-foot contour interval derived from the DTM model.

c. *Cover and control sheet:* The first sheet shall be a cover sheet showing the control sketch, survey control tabulation, sketch layout or index, legend, project location map, survey notes, north arrow, graphic scale, grid ticks, and large signature block. Tabulate, plot, and list the horizontal control used for the survey on the final drawing.

d. *Plan sheets:* The plan sheets shall be prepared to a scale of $1'' = 100'$, showing notes, title block, grid, north arrow, graphic scale, legend, sheet index, and file number. Sheets shall be oriented with north to the top. The extreme right 7 inches of the sheet shall be left blank for notes, legends, and so forth. The second sheet and all sheets following shall be a continuation sheet and shall have a minimum of two notes, note 1: See Drawing number 1 for notes, note: Refer to Survey No. 99–267.

6. *Map content:*

   a. *Coordinate grid (NAD 83):* Grid ticks of the applicable State Plane Coordinate System shall be properly annotated at the top, bottom, and both sides of each sheet. Spacing of the ticks shall be five (5) inches apart.

   b. *Control:* All horizontal and vertical ground control monuments shall be shown on the maps in plan and tabulated.

   c. *Topography:* The map shall contain all representable and specified topographic features that are visible or identifiable.

   d. *Spot elevations:* Spot elevations shall be shown on the maps in proper position.

   e. *Map edit:* All names, labels, notes, and map information shall be checked for accuracy and completeness.

   f. *Map accuracy:* All mapping shall conform to the National Map Accuracy Standards except that no dashed contour line will be accepted.

7. *Deliveries:* On completion, all data required shall be delivered or mailed to Design Branch, Survey Section at the address shown in the contract, and shall be accompanied by a properly numbered, dated, and signed letter or shipping form, in duplicate, listing the materials being transmitted. All costs of deliveries shall be borne by the contractor. Items to be delivered include, but are not limited to the following:

   a. GNSS network plan, (before GNSS work commences).

   b. GNSS raw data along with field observation log sheets filled out in field with all information and sketches.

   c. Computation files.

   d. Field books.

   e. X, Y, Z, and descriptor ASCII file for each beach profile and one of all beach profile data merged together.

   f. DTM file.

   g. Sheet files at 1" = 100'.

   h. Excel file with Site ID, X, Y, Z, and Azimuth of profile line.

(Source: Adapted from USACE EM 1110–1–1003, 12–9–12–13.)

11.5 TOPOGRAPHIC SURVEY COST ESTIMATING

11.5.1 SAMPLE COST ESTIMATE FOR TOPOGRAPHIC SURVEY

The following cost computations are representative of the procedures used in preparing the contract price schedule in Table 11.3. Costs and overhead percentages are shown for illustration only.

1. *Labor:* Labor rates are direct costs and their estimates can be obtained from a number of sources, such as prior contract rates, trade publications, and Department of Labor published rates.

2. *Indirect overhead costs:* Overhead is an indirect cost—a cost that cannot be directly identified but is necessary for the normal operation of a business. Overhead is normally broken into two parts: direct and G&A. Direct overhead includes items such as benefits, health plans, retirement plans, and life insurance. G&A includes office supervision staff, marketing, training, depreciation, taxes, insurance, utilities, communications, accounting, and downtime. (Care must be taken to ensure there is no duplication between G&A overhead costs and direct costs. An example of duplication might be a maintenance contract for a total station being included in both G&A and directly on the equipment cost.) Usually, direct and G&A overheads are combined into one amount and applied as a percentage against the base labor cost.

Table 11.5 shows a sample labor rate computation for two selected line items: a Party Chief and a Survey Aid (2,087 hours per year assumed). Direct and G&A overheads are broken out for the Party Chief but are shown combined for the Survey Aid. Daily rate, hourly rate, and overtime rate are shown.

3. *Estimating equipment and instrumentation costs:* Table 11.6 shows an example of instrumentation cost estimates. Total station and GNSS rates

TABLE 11.5
Sample Labor Rate Computations

Supervisory survey tech (party chief)	$42,776.00/year
Overhead on direct labor (36%)	$15,399.36/year
G&A overhead (115%)	$49,192.40/year
Total:	$107,367.76/year or $411.57/day or $51.44/hour
	Overtime rate: $42,776/2,087 × 1.5 = $30.74
Survey aid	$23,332.00/year
@151% O/H (36%+115%)	$58,563.32 or $224.49/day or $28.06/hour
	Overtime rate: $23,332/2,087 × 1.5 = $16.77

TABLE 11.6
Estimating Survey Instrumentation and Equipment Costs

Total Station: Data Collector, Prisms, and so forth

$32,000 purchase cost @ 5 years life @ 120 day/year utilization			$53/day
A lease rate of $600/months @ 10 days utilization/months			$60/day
RTK topographic system: 2 geodetic GNSS receivers, batteries, tripods, data≈collectors, and so forth			
$30,000 purchase cost @ 4 years @ 100 day/year			$75/day
Laptop, field—with COGO, GNSS, and CADD software $15,000 purchase cost @ 3 years @ 200 day/year			$25/day
Survey Vehicle: $50,000 @ 4 years @ 225 day/year		$55/day	
plus O&M, fuel, and so forth		$25/day	$80/day
(A purchase or lease rate may be used. Vehicle costs should not include the items covered under G&A, such as liability insurance)			
Misc. materials (field books, survey supplies, etc.)			$15/day

used are approximate (2004) costs. Associated costs such as insurance, maintenance contracts, and interests are presumed to be indirectly factored into G&A overhead. If not, then such costs can be directly added to the basic equipment depreciation rates. Other equally acceptable methods for developing daily costs of equipment can be used. The major variables in estimating costs are as follows:

a. *Utilization rates:* A particular survey instrument may be used (charged) only a limited number of days in a year. A total station or vehicle may be utilized well over 200 days a year, whereas other instruments are not used on every project. For example, a $150,000 terrestrial scanner may be used only 20 days a year. If the annual operating cost of this instrument (without operator) is, say, $40,000, then the daily rate is $2,000/day. This is the amount the contractor will charge to recoup purchase or lease expenses (not including profit). A two-man survey crew may carry both a total station and GNSS system with them in the field. Even though only one of these systems can be used on a given day, both systems are chargeable for utilization. Utilization rates are difficult to estimate—they can widely vary from contractor to contractor and with the type of equipment.

b. *Equipment cost basis:* There are a number of methods to estimate the cost basis of a particular instrument. Trade publications (e.g., *POB, Professional Surveyor,* and *American Surveyor*) contain tabulations and advertisements with purchase costs, loan costs, rental costs, or lease costs. If an item is purchased, then an estimated life must be established—usually varying between 3 and 7 years for most electronic equipment and computers. Assuming the instrument is purchased on a loan basis, the annual/monthly cost can be estimated, for example, a $40,000 instrument purchased over 5 years at 5% is $755/month. At 15 days/month estimated utilization, the daily rate would be $50/day. Lease rates published in trade publications also provide estimates for costs. Rental rates are also applicable. In general, rental rates could run between 5% and 15% of the original purchase cost, per month. Thus, a $40,000 instrument could be rented for $4,000 per month, assuming a 10% rate. If it is utilized 15 days a month, then the daily rental rate would be $266/day.

4. *Travel and per diem:* Travel and per diem costs are usually based on the geographical location, transportation mode (air, land, sea), mob/demob times, and so forth.

5. *Material costs:* The material cost estimate in Table 11.6 could have been easily included in G&A overhead, given that they are usually small amounts relative to labor and equipment line items.

6. *Combined crew rates:* The labor and equipment line items described earlier can be combined to obtain a single rate for a one- or two-man topographic crew; for example, a one-man crew of a Party Chief, total station, computer, vehicle, and miscellaneous supplies. Using the rates in Tables 11.5 and 11.6, the daily cost of a one-man crew would be computed as shown in Table 11.7.

TABLE 11.7

Sample Computation for One-Man Topographic Crew

Supervisory survey tech (party chief)—(includes 151% O/H)	$411.57/day
Total station (robotic): data collector, prisms, and so forth	$53.00/day
Vehicle	$15.00/day
Miscellaneous expenses	$80.00/day
Subtotal	$559.57/day
Profit @ 10%	$55.96/day
Total crew rate	$615.53/day

Note: Travel and per diem expenses would be added separately.

11.5.2 SAMPLE SCOPE OF WORK FOR TOPOGRAPHIC SURVEY

The following is an example of a letter request for proposal for topographic surveying services. Included with the letter is the sample scope of work.

SAMPLE LETTER REQUEST FOR PROPOSAL

26 March 2002
Surveying and Mapping Section
EarthData International
45 West Watkins Mill Road
Gaithersburg, MD 20878
SUBJECT: Contract No. DACW27-00-D-0017
Gentlemen:

Enclosed is a scope of work dated March 25, 2002 for topographic mapping and boundary survey of a proposed site in the vicinity of Cleveland, OH. This work is for a delivery order under the above-referenced contract. Please submit your proposal no later than ten (10) calendar days after receipt of this letter. Return your proposal by mail or by fax to 502/315–6197. Mark your proposal to the ATTENTION OF CELRL-CT (PP&C 20946148).

For technical questions concerning the scope of work contact Chris Heintz at 502-345-6508.

Sincerely,
Enclosure
CF:
CELRL-ED-M-SM (C. Heintz)
(Source: Adapted from USACE EM 1110–1–1005, 13–11.)

SAMPLE SCOPE OF WORK

Contract No.DACW27-00-D-0017
EarthData International
Date: March 25, 2002
Project: Topographic Mapping and Boundary Survey

GENERAL

The contractor shall provide all labor, material, and equipment necessary to perform necessary professional surveying and mapping for the Louisville District. The work required consists of gathering field data, compiling this data into a three-dimensional digital topographic map of the proposed site.

This project also requires performing a boundary survey of the site. The details of the boundary survey are described in the attached scope of work.

The contractor shall furnish the required personnel, equipment, instrumentation, and transportation as necessary to accomplish the required services and furnish digital terrain data, control data forms, office computations, reports, and other data with supporting material developed during the field data acquisition and compilation process. During the prosecution of the work, the contractor shall provide adequate professional qualification and quality control to assure the accuracy, quality, completeness, and progress of the work.

TECHNICAL CRITERIA AND STANDARDS

The following standards are referenced in specification and shall apply to this contract:

USACE EM 1110–1–1005, Topographic Surveying.

USACE EM 1110–1–1002, Survey Markers and Monumentation.

ASPRS accuracy standards for large-scale maps. Digital Elevation Model Technologies and Applications: The DEM User's Manual.

SCOPE OF WORK

Professional surveying, mapping and related services to be performed are defined below. Unless otherwise indicated in this contract, each required service shall include field-to-finish effort. All mapping work will be performed using appropriate instrumentation and procedures for establishing control, field data acquisition, and compilation in accordance with the functional accuracy requirements.

Three-dimensional digital topographic map will be compiled in meters at a scale of 1:600, with 0.25 m contours. The mapping area is outlined on the attached map. All planimetric features will be shown. This includes, but is not limited to, buildings, sidewalks, roadways, parking areas (including type such as gravel, paved, and concrete), visible utilities, trees, road culverts (including type and size), sanitary manholes, storm manholes, inlets and catch basins, location of fire hydrants and water valves, location and type of fences and walls.

A referenced baseline with a minimum of two points will be established adjacent to each site. The location of the baseline will be set in an area that will not be disturbed. At least two benchmarks will be set within the map area. The baseline stations and benchmarks will be referenced and described. In addition to showing the descriptions in a digital file, a hard copy of the descriptions will be submitted with the project report.

The coordinates of the mapping projects will be tied to the local State Plane Coordinate System NAD83 and vertically tied to NAVD 1988.

PROJECT DELIVERIES

The contractor will submit the final topographic map in digital format (AutoCAD format *.dwg or *.dwt on CD-ROM or DVD). The digital file will be created in 3D with the topographic and planimetric elements placed at their actual X and Y coordinate locations. The global origin will be 0,0 and current CADD standards will be applied—symbology, layers, colors, line weights, and so forth.

A project report will be compiled. This report will contain a general statement of the project, existing geodetic control used to establish new monumentation, condition of existing monuments, baseline and BM descriptions and references, adjustments, procedures and equipment used, any special features unique to the project, and personnel performing the surveying and mapping.

All field notes will be submitted in a standard bound survey field book or if electronic data collection methods were employed, all digital raw data files, in ASCII format will be submitted. If electronic data collection was the method of choice for capturing the information, the final X, Y, and Z coordinate file, in ASCII format, will be submitted with the raw data file.

A metadata file describing the project.

QUALITY CONTROL

A quality control plan will be developed and submitted. The quality control plan will describe activities taken to ensure the overall quality of the project.

The accuracy of the mapping will meet or exceed ASPRS map accuracy class II.

Map verification will be performed at each site. The verification will be accomplished by collecting coordinates for ten random points at each site and comparing them with the coordinates of the same points on the finished map. The random points will not be used to compile the finished map. Differences between the field-test information and the finished map will be compared with differences allowed by ASPRS map accuracy class II standards. Any areas found to be out of compliance must be corrected before submittal. A summary of the actual versus allowable differences along with a statement that mapping meets ASPRS map accuracy class II standards will be provided with the data.

SCHEDULE

All work will be completed and submitted by 15 May 2002.
(Source: Adapted from USACE EM 1110–1–1005, 13-12–13–14.)

11.6   AERIAL SURVEY AND MAPPING COST ESTIMATING

11.6.1   Cost Phases of Mapping Operation

This section presents the elements that should be addressed when costing for a digital photogrammetric mapping project. The costing procedures are only representative and can be used to estimate all or only certain parts of a mapping project. Initially, it is important to design the various photogrammetric mapping procedures together in a logical sequence (i.e., depicting a typical photogrammetric mapping and orthophoto production flow). Cost estimates should account for all the significant cost phases including aerial photography, ground control, aerial triangulation, and DTM development. Here are examples of how production hours can be estimated:

1. *Calculating production hours for aerial photography*
   Direct Labor:
   Flight preparation = _____ hours
   Take off/landing = _____ hours
   Cross-country flight = _____ hours
   Photo flight = _____ hours
   Develop film = _____ photos = _____ hours
   Check film = _____ photos = _____ hours
   Title film = _____ photos = _____ hours
   Contact prints = _____ photos = _____ hours
   Equipment rental:
   Aircraft = project mission hours = _____ hours
   Airborne GNSS = project mission hours = _____ hours
   (if not included in aircraft rental)
   Film processor = develop film hours = _____ hours
   Film titler = title film hours = _____ hours
   Contact printer = contact print hours = _____ hours

2. *Cost items for photo control surveying*
   Airborne GNSS control is commonplace. However, if ground control methods are needed or preferred, the following items are generally considered in the calculation of costs associated with photo control surveying:
   a. Distance from survey office to site
   b. Distance to horizontal reference
   c. Distance to vertical reference
   d. Time to complete horizontal photo control or number of points required
   e. Time to complete vertical photo control or number of points required.

3. *Aerotriangulation*
   Direct labor:
   Photo scan = _____ photos = _____ hours
   Model orientation = _____ models = _____ hours
   Coordinate readings = _____ photos = _____ hours
   Computations = _____ models = _____ hours
   Equipment rental:
   Scanner = scanning hours = _____ hours
   Workstation = aerial triangulation hours = _____ hours
   Computer = computations hours = _____ hours

4. *Compilation and digital mapping*
   The following items are to be calculated, estimated, or measured to assist in computing costs associated with digital mapping:
   a. Number of stereomodels to orient
   b. Number of acres or stereomodels to map
   c. Complexity of terrain character
   d. Format translations of digital data.

Production Hours for Stereomapping
Model Setup:
Model orientation = _____ models = _____ hours
Photo scan = _____ photos = _____ hours
(if not done previously)
Digital data capture:
Planimetric features—The planimetric feature detail in each of the models should be assessed based on the amount and density of planimetric detail to be captured in each stereomodel. Highly urban area stereomodels will require more time to compile than rural area stereomodels.

Topography—Topographic detail must consider the character of the land to be depicted. For example, 1-foot contour development in a relatively flat terrain requires much less time than collection of 1-foot contours in mountainous terrain.

5. *Orthophoto images*

Current technology allows for total softcopy generation of orthophotos. If a softcopy stereo compilation is used for the digital elevation model, the same scanned images may be used to generate orthophotos. However, if the DTM is collected with analytical stereoplotter and diapositives created, then a clean set of diapositives must be made and scanned for orthophoto generation. Assume one method or the other in developing a cost estimate.

Production Labor

	Hours	Unit Cost	Total Cost
Aerial photography			
Aerotriangulation			
Model setup			
Planimetry			
Topography			
Orthophotography			
Total			

Direct Costs

	Hours	Unit Cost	Total Cost
Film			
Prints			
Diapositives			
CDs, disks, or tapes			
Aircaft w/camera			
Stereo plotter			
Workstation			
Scanner			
Total			

A summary of the itemized production hours is shown in the following list. Current unit costs should be established for each task to be used in a project. The unit costs should include necessary equipment and labor.

11.6.2   SAMPLE SCOPE OF WORK

The following is a sample scope of work for photogrammetric mapping.

SAMPLE PROJECT #1
1. *Description of work:*
    Mapping of portions of the ALCOA site has been requested. The area to be mapped is approximately 800 acres. The final mapping products requested are digital, planimetric, and topographic in ARC/INFO format. The map scale will be 1 in. = 50 ft with 1 in. contours. The aerial photography will be flown at a negative scale of 1 in. = 330 ft utilizing panchromatic (black and white) film. Minimal ground survey control to perform aerotriangulation, develop DTMs, and produce the digital mapping will also be obtained. All photography will be flown at approximately 1,980 ft above mean terrain. The final mapping will fully comply with ASPRS class I Accuracy Standards for mapping at a horizontal scale of 1 in. = 50 ft with a DTM suitable for generation of 1 ft contours.

2. Contractor shall provide equipment, supplies, facilities, and personnel to accomplish the following work.
    a.   Contractor will establish an aerial photo mission and ground survey control network for the project. Photography will be flown with 60% forward lap and approximately 30% side lap. GNSS data collection and processing will include latitude, longitude, and ellipsoid elevation for each photo center. All ground survey plans including survey network layout, benchmark to be used, and so forth shall be approved prior to initiation of project. The plan submitted shall include but not limited to maps indicating proposed GNSS network, benchmarks to be used, flight lines, and project area.
    b.   Additional ground survey data will be collected to be used in the mapping process and to check the final mapping. All original notes for the surveys shall be submitted and all survey data shall be in the local State Plane Coordinate System, referenced to WGS-84. Vertical datum will be NAVD 88 adjustment.
    c.   Two sets of contact prints will be made in accordance with the technical specifications provided in the contract. One set of prints will be used as control photos for mapping. The control prints will have all ground control marked on the back and front of each photo.
    d.   Ground control will be utilized to perform analytical aerotriangulation to generate sufficient photo control points to meet National Map

Accuracy Standards for mapping at a horizontal scale of 1 in. = 50 ft with a DTM suitable for generation of 1-ft contours. The contractor will produce a written report discussing the aerotriangulation procedures used, number of ground control points used, any problems (and how they were solved), the final horizontal and vertical RMSE, and how to read the aerotriangulation print out (units, etc.). The written report will be signed and dated by the author.

e. The 1 in. = 330 ft photo diapositives will be utilized, and planimetric feature detail (all that can be seen and plotted from the photography) and DTM data will be collected for topographic mapping at a horizontal scale of 1 in. = 50 ft with 1 ft contours. DTM production will utilize collection of mass points and breaklines to define abrupt changes in elevation. Data will be delivered in a suitable digital Geographic Information System (GIS) format on a disk.

f. The contractor will produce the planimetric feature data, DTM, and contour files in a digital format (e.g., AutoCAD *.dwt) on a disk.

g. The contractor will provide metadata for the aerial flight, ground control, and mapping data sets in accordance with the applicable provisions of the Content Standards for Digital Geographic Metadata by the Federal Geographic Data Committee.

3. *Delivery items:*

a. Copy of computer printout of aerotriangulation solution, and one copy of written aerotriangulation report.

b. Copy of camera calibration reports.

c. One copy of digital planimetric feature files and topographic data files at a horizontal scale of 1 in. = 50 ft, with 1 ft contours. One copy of the DTM files suitable for 1 ft contours.

d. All survey data (including ground surveys), raw GNSS files, any other survey information developed and/or collected for the project.

e. Two sets of panchromatic (black and white) prints and one set of diapositives.

f. Flight line index on paper maps indicating the flight lines and beginning and ending frames for each flight line along with altitude and scale of the photography.

g. Metadata on CD-ROM for aerial photography, ground control, and mapping data sets.

4. Schedule and Submittal:

a. The contractor will capture the photography before November 30, 1998. The contractor will deliver all final products (including digital data files) within 45 calendar days after photography is flown.

b. All material shall be delivered at the contractor's expense.

(Source: Adapted from USACE EM 1110–1–1000, 4–34–4–38.)

11.6.3 Sample Cost Estimate

SAMPLE COST ESTIMATE WORKSHEET
  PHOTOGRAMMETRIC MAPPING
  CONTRACT NUMBER

Cost Item	Units	Rate	Amount
Project manager	24 hours	$30	$720.00
Chief photogrammetrist	20 hours	$30	$600.00
Photogrammetrist	40 hours	$23	$920.00
Aerial pilot	3.5 hours	$19	$66.50
Aerial photographer	3.5 hours	$16	$56.00
Computer operator	40 hours	$23	$920.00
Compiler	401 hours	$15	$6,015.00
Drafter/CADD operator	109 hours	$11	$1,199.00
Photo lab Technician	8 hours	$9	$72.00
Total direct Labor			$10,568.50
Combined overhead on direct labor and G&A overhead at 160.5%			$16,962.44
Total direct labor and overhead			$27,530.94
Direct costs			
Airplane w/camera and GNSS	2 hours	$700.00	$1,400.00
B/W prints	80	$0.55	$44.00
B/W diapositives	40	$1.65	$66.00
CD-ROM	2	$5.00	$10.00
Total direct Costs			$1,520.00
Total direct Labor, overhead and direct costs			$29,050.94
Profit @ 12%			$3,486.11
Subcontract			
Ground surveys			
18 H/V, 3 field Days plus			
1 day Computations			$6,400.00
Total			$38,937.05

BIBLIOGRAPHY

U.S. Army Corps of Engineers (USACE). 2002a. *Engineering and Design Manual for Photogrammetric Mapping*. EM 1110–1–1000. Washington, DC: Department of the Army.

U.S. Army Corps of Engineers (USACE). 2002b. *Engineering and Design Manual for Geodetic and Control Surveying*. EM 1110–1–1004. Washington, DC: Department of the Army.

U.S. Army Corps of Engineers (USACE). 2003. *Engineering and Design Manual for NAVSTAR Global Positioning System*. EM 1110–1–1003. Washington, DC: Department of the Army.

U.S. Army Corps of Engineers (USACE). 2007. *Engineering and Design Manual for Control and Topographic Surveying*. EM 1110–1–1005. Washington, DC: Department of the Army.

EXERCISES

1. Carry out a research and develop a list of current labor rates for costing services offered by surveying and geomatics professionals (e.g., geodesists, professional land surveyors, survey technicians, Global Positioning System (GPS)/GNSS analysts, photogrammetrists, photogrammetry technicians, survey field assistants, survey party chief, GIS professionals, etc.). (Possible sources of information include trade publications and employment statistics from the Department of Labor or a similar organization.)

2. Using current market rates, develop a list of unit prices (i.e., daily rates) for equipment rental or purchase, for the following (for each unit price indicate whether it is a rental or purchase costing):
   a. Geodetic quality GPS/GNSS receiver
   b. Robotic total station
   c. Digital level
   d. Laser scanner
   e. LIDAR aircraft
   f. Vehicle

3. Using some of the information from answers to the previous questions, develop a cost estimate for a one-man GNSS survey crew involving a licensed professional surveyor, a geodetic GNSS receiver, one vehicle, and materials. Explain all assumptions, such as overhead and profit costs.

4. How is a cost estimate different from a cost budget?
   (Hint: See Chapter 1.)

5. Develop a cost budget using a sample request for proposal (outlining the scope of work) or one in your recent experience.

12 Writing Geomatics Proposals

12.1 INTRODUCTION

This chapter is concerned with the writing of geomatics proposals, which is a process for contracting surveying and geomatics engineering services. Proposals can be classified as either small/large, government/industry, or formal/informal. We will assume the context of writing a formal proposal and the premise of the chapter is that the basic process is the same for a small contract or a major project worth millions of dollars; what changes is the scale.

First, it is important to understand where proposals fit in the scheme of contracting processes and procedures. As an example, and on the basis of the assumptions made in Chapter 11, the following paragraphs summarize a typical contracting process under the Brooks Architect-Engineer Act, Public Law 92–582 (10 US Code 541–544):

1. *Announcements for contracting services:* Requirements for services are publicly announced and firms are given at least 30 days to respond to the announcement. The public announcement contains a brief description of the project, the scope of the required services, the selection criteria, submission instructions, and a point-of-contact.
2. *Selection criteria:* The client or client's representative set the criteria for evaluating prospective contractors. Such criteria may be listed in their order of importance and that order may be modified based on specific project requirements. For instance, the criteria list may include
   1. Professional qualifications necessary for satisfactory performance
   2. Specialized experience and technical competence in the type of work required
   3. Past performance on similar contracts, for example, in terms of cost control, quality of work, and compliance with performance schedules
   4. Capacity to perform the work in the required time
   5. Knowledge of the locality of the project
   6. Geographic proximity.
3. *Selection process:* The evaluation of firms is conducted by the client or a formally constituted Selection Board or Committee. The board or committee is made up of highly qualified professionals having experience in the services being contracted. The board or committee evaluates each of the firm's qualifications based on the advertised selection criteria and develops a list of the most highly qualified firms for a single award. As part of the evaluation process, the board conducts interviews (e.g., by telephone) with

these top firms prior to ranking them. The firms are asked questions about their experience, capabilities, organization, equipment, quality management procedures, and approach to the project. The top firms are ranked and the selection is approved by the designated selection authority. The top ranked firm is notified that it is under consideration for the contract. Unsuccessful firms are also notified, and are afforded a debriefing as to why they were not selected, if they so request.

4. *Negotiations and award:* The highest qualified firm ranked by the selection board is provided with a detailed scope of work for the project, project information, and other related technical criteria, and is requested to submit a price proposal for performing the work. Once a fair and reasonable price is negotiated, the contract is awarded.

A written proposal in response to the announcement for contracting services is the basis upon which the selection process is conducted. However, it should be noted that in the aforementioned case of Brooks Act, cost or pricing is not considered during the selection process. A fair and reasonable price is negotiated with the highest qualified firm, only after the selection process is completed. This procedure is typical in the qualifications-based selection (QBS) process as defined by the Brooks Act. Entities or organizations not bound by the Brooks Act are likely to have different criteria for evaluating proposals.

In practice, some clients might require that a full proposal be submitted with a separately sealed cost proposal (so that only the cost proposal of the winner would be opened by the client). Here, for the sake of discussing a complete proposal layout, we will assume further that there are no requirements (e.g., by the Brooks Act) to exclude the cost from the initial response to a request for proposal (RFP).

As seen in Chapter 1, a proposal is an estimate of the time and cost required to complete a project according to the scope of required services. Writing the proposal is itself a project because it requires a series of efforts within the constraints of time. The efforts include those preceding the actual writing of the proposal. For instance, you must first gather preliminary intelligence and make a bid decision. This will be based on understanding of project requirements and assessment of your resources and capabilities to meet those requirements, compared with those of your competitors.

12.2   HOW TO WRITE A WINNING PROPOSAL

A good proposal should include

1. Evidence of clear understanding of the client's problems
2. An approach and program plan or design that appears to the client well suited to solving the problem and likely to produce the results desired
3. Convincing evidence of qualifications and capability for carrying out the plan properly
4. Convincing evidence of dependability as a consultant or contractor
5. A compelling reason for the client to select the proposal, that is, a winning strategy.

"A winning proposal is a project blueprint and, therefore, specific. Problems during the project concerning budget, schedule, and scope of work often relate to deficiencies in the proposal" (Stasiowski, 2003, 4).

A winning proposal focuses on the client's needs and concerns, not those of the proposer. You need to know your potential clients very well in order to identify their real issues (not always those stated in the RFP), and counter those with the benefits you provide that will solve their problems.

A winning proposal will show that you understand the big picture—your client's overall business and the competitive world in which they operate. It will show how you will add value to their operation.

The content presents clearly your logical, technical approach to their project, and shows that your plan will work. It presents an attractive fee in relation to the value you will add. It shows a clear link between the elements of the proposal and the evaluation criteria.

A well-written, winning proposal demonstrates your understanding of the client's needs and desires. It restates the criteria for selection and details how your firm meets those criteria. It demonstrates the competencies and experience that your project team brings to this specific project. It is readable, concise, attractive, and well organized.

Losing proposals tend to be too generic: they don't reflect the client's specific needs, wants, or concerns. They focus on the features offered by the proposing contractor or firm instead of the benefits those features could provide to the client. They are often packed with statistics and exhibits that are not relevant to the project, or they are padded with good-looking boilerplates that say little or nothing relating to the client's concerns. Losing proposals are disorganized, hard to read, incomplete, too long, or too short.

12.3 LAYOUT OF THE PROPOSAL

So, what is the layout of the proposal supposed to look like? What things should be included in the proposal format?

If there is no written RFP, or if the written RFP does not specify outline or format, then there are no rules for the layout and design of your proposal. The only standard to apply to the proposal's appearance is whether it fulfills the proposal evaluator's expectations. If they haven't told you what they are or written them into an RFP, then all you can do is make your proposal legible. Your proposal layout should be highly readable and make it easy to locate information. You should make extensive use of graphics, because they enhance the readability of the document and convey information well. In the absence of instructions to the contrary, your headings, typefaces, margins, headers/footers, and other formatting attributes can be anything that you want that achieves the goal of your proposal.

The following information is a general format adapted from *Sample Proposal Format for RFP Responses* (MoreBusiness.com, May 17, 1999). The content changes based on what an RFP requires so you can fill in after figuring out how you will solve the client's requirements:

1. *Background:* Briefly go over the general requirements of the client. Example: AA Consultants would like land surveying services for a record of survey of the highway 111 Corridor within the City of Palm Desert, California.

2. *Scope:* Discuss in detail each item in the RFP and how you intend to tackle it. Use diagrams to illustrate your configuration. This will be the longest section of your proposal and will probably have several subsections.

3. *Schedule:* When do you anticipate starting? How long will each task take? Make a table of your expected schedule for completing the project.

4. *Personnel:* This is an optional section. Some firms like to see who will be working on the project. This is more important for government projects. Put the bios and resumes here.

5. *Cost:* Breakdown the cost by equipment and personnel time to come up with your expected budget. Include payment terms, discounts for early payment, and other cost or payment information.

6. *Supporting information:* Add any supporting info here (e.g., if you're trying to convince them to use a specific type of technology, back up your reasoning with third-party quotes, research, test results, etc.). You can also add information about similar projects you have completed for other firms and what the results were of those. Include testimonials from clients, clippings from newspapers, and so forth.

12.4   MANAGING THE WRITING PROCESS

Many firms consider proposal writing a necessary unpleasantness, rather than part of an ongoing, dynamic, well-informed marketing effort. They prepare proposals on an ad hoc basis, in preference to maintaining an efficient, thoughtful preparation process. Often they go after lots of RFPs, even if the work isn't what they prefer to be doing, or they go after appropriate projects even when they know they won't get the work. They bulk up the proposal with boilerplate, unnecessary information, or untargeted writing. Then, they don't get the work. This further convinces them that proposal-writing is a necessary unpleasantness. (Stasiowski, 2003, 6)

It doesn't have to be so!

12.4.1   GETTING STARTED

First, it is necessary to create a preliminary proposal writing timeline. The process can be more or less complicated, and usually involve a team of several people or only a few, depending on the firm and its background.

12.4.2   ASSIGNING INDIVIDUAL AND TEAM RESPONSIBILITIES

The team should start by dividing the proposal writing responsibilities. They should outline who is responsible for each portion of the proposal, each with a strict time-line. This process should culminate into individual and team responsibilities as outlined in Section 3.3 of the book. It is necessary to have a team leader to ensure that the team is meeting deadlines and objectives that will result in a quality proposal.

12.4.3 Outlining the Process

The team members should completely understand the RFP's overall objective, the information that needs to be addressed, and be able to define the strategic approach before they start writing the proposal. This process helps the team identify information gaps and delineates the resources responsible for each portion of the proposal.

Three essential steps are necessary to produce a stand-out proposal:

1. Evaluate whether you should spend your time and money proposing on the project. Consider your strategic plan, your financial objectives, and the value of the prospective client to your business.
2. Look at the project from your client's point of view. Which issues do they really care about? How can you address these issues? How can you prove that you can do it?
3. Have a review team (members of your firm who are not involved in writing the proposal) pretend to be the client and give the proposal draft a critical analysis from that standpoint.

Whether you have weeks or only days to complete the proposal, these three steps should be part of the process. Adjust the scale of your proposal efforts accordingly.

12.4.4 Incorporating Revisions and Feedback

To put the finishing touches on the proposal, someone should be in charge of editing and reviewing the proposal including all feedback from the review team. Ensure that the proposal meets all the criteria outlined in the RFP.

12.4.5 Submitting the Proposal

Companies, universities, and organizations may write a winning proposal, only to have it rejected because a form was not submitted; bios were not in the correct format; or five case studies, instead of three, were included. These seemingly small issues can be the difference between winning and losing a contract. The proposal team should ensure that all the i's are dotted and the t's are crossed and that all required items are properly included.

12.4.6 Things to Avoid

According to the article "Proposal Writing for Government Contracting" (www.fedmarket.com/articles/government-proposal-writing.shtml) by Richard White, the following statements caused proposals to be rejected by the state of California. Do your best to avoid such statements.

1. A bid stated, "The prices stated within are for your information only and are subject to change."
2. A bid stated, "This proposal shall expire thirty (30) days from this date unless extended in writing by the xyz Company." (In this instance, the award was scheduled to be approximately 45 days after bid submittal date.)

3. A bid for lease of equipment contained lease plans of a duration shorter than that which had been requested.

4. A personal services contract stated, "xyz, in its judgment, believes that the schedules set by the State are extremely optimistic and probably unobtainable. Nevertheless, xyz will exercise its best efforts..."

5. A bid stated, "This proposal is not intended to be of a contractual nature."

6. A bid contained the notation, "Prices are subject to change without notice."

7. A bid was received for the purchase of equipment with unacceptable modifications to the purchase contract.

8. A bid for lease of equipment contained lease plans of a duration longer than that which had been requested in the invitation for bid (IFB) with no provision for earlier termination of the contract.

9. A bid for lease of equipment stated, "This proposal is preliminary only and the order, when issued, shall constitute the only legally binding commitment of the parties."

10. A bid was delivered to the wrong office.

11. A bid was delivered after the date and time specified in the IFB.

12. An IFB required the delivery of a performance bond covering 25% of the proposed contract amount. The bid offered a performance bond to cover X dollars, which was less than the required 25% of the proposed contract amount.

13. A bid did not meet the contract goal for participation (e.g., EEO) and did not follow the steps required by the bid to achieve a "good faith effort."

12.5   SAMPLE PROPOSAL FORMAT

In this section, we look at a sample proposal format with illustrations. Be sure to keep in mind, though, that this is only a hypothetical proposal format, an example. It is not an absolute formula to be followed blindly.

Most proposals will start with a cover letter, title page, executive or client summary, and a cost summary or estimate, and then will include additional material as needed. The proposal should be of high quality in terms of technical content and writing.

12.5.1   COVER LETTER

SAMPLE GENERIC TEMPLATE
   (*Date*)
   (*Client*)
   (*Address*)
   Dear (*Client*):
   PROPOSAL FOR (*Project Title/Type of Services*)
   (*Proposal Number, if applicable*)
   (*Briefly describe your strategy and why you are submitting the proposal.*)
   (*Briefly state your unique selling proposition that solves the client's problem and optionally attract them with anything extra you think of.*)

(Briefly reference RFP and project name, if needed state how long the proposal is valid for and include any additional information that the client may need to know immediately.)

(Invite the client to contact you should they have any questions or need additional information.)

Sincerely,

(Name of Proposer)

(Title of Proposer)

(Contact Details)

Enclosures

The following is a sample cover letter for a cost estimate (proposal) transmittal. It is brief, to the point, and does not follow the previous generic template in every detail. For example, the proposer is not interested in describing a unique selling proposition or strategy. This might apply, for instance, if the proposer has already been selected by the client, and the proposal only serves the purpose of initiating a price negotiation process.

SAMPLE LETTER #1

(Date)

(Client)

(Address)

Dear *(Client)*:

PROPOSAL FOR PROFESSIONAL SURVEYING & ENGINEERING DESIGN FOR REGRADING OF EXISTING AIRSTRIP AT *(Insert Project Location)*

PROPOSAL # 7–10–1–20101

We are pleased to submit this proposal for professional surveying and engineering design services for the above referenced project. The following appendices will describe our proposal.

Appendix I—Scope of Work

Appendix II—Fees for Professional Services

Proposer, Inc. appreciates the opportunity to submit this proposal for your review and acceptance and looks forward to performing this work for you.

If you should have any questions or need additional information, please call.

Sincerely,

PROPOSER, INC.

(Signature)

(Name and Title of Proposer)

(Contact Details)

Enclosures

In the following sample letter, the proposer is responding to an informal RFP. The prospective client could have said something like "send me a proposal," after a series of talks or conversations with the consultant. In such a case, it is important to talk through every phase of the project with the soon-to-be client before writing the proposal. Questions must be asked until the scope, budget, and duration of the project are crystal clear.

SAMPLE LETTER #2

(Date)

(Client)
(Address)
Dear *(Client)*:
PROPOSAL FOR MAPPING FOR LAND DEVELOPMENT
PROPOSAL # 7–10–1–20102
I enjoyed meeting with you on *(Date)* to discuss the possibility of mapping your land for subdivision. As you well know, the profits on grapes have been remarkably low for the last few years, with no hope for future gains in the global market. ABC County is allowing so many acres of agricultural land to be developed each year in an effort to boost the local economy. The 30 acres that you own at *(Location/Address)* is a prime location for development.

ABC Surveyors and Land Developers, Inc. started business 5 years ago in the ABC Valley. We specialize in helping clients develop their land into housing tracts. We are proud of the service that we provide, and we have been very successful with it. We have now helped map and develop over 200 acres of land in the Valley, all with excellent results.

Please be aware that developing your land without the services of a licensed professional could cost you time and money. That is where we come in. Our experience in the county, our knowledge of permit and license requirements, and our proven ability to contract all phases of mapping for development, including title search through completion, in a timely fashion make us the obvious choice for a land development partner. So far, every mapping project we have undertaken has been accomplished on time, and generally on budget. We pride ourselves on hard work and not only our ability to get the job done, but the ability to get it done right.

I have attached a proposal for you to browse through. It is by no means all inclusive, but it should give you a better idea what our vision is for the costs and benefits involved. I look forward to meeting with you again.

Sincerely,
ABC Surveyors and Land Developers, Inc.
(Signature)
(Name and Title of Proposer)
(Contact Details)

12.5.2 TITLE PAGE

(Proposer's Business Title/Name of Firm)
(Address)
(Telephone Number)
(Fax Number)
(Web Site)
(Project Title)
    Prepared for: *(Name of Client)*
(Title, Client's Business Name)
    Prepared by: *(Proposer's Name)*
(Title, Proposer's Business Name)
    *(Motto)**
    (Proposal Number, if applicable)

[* A motto can be something like "ABC will meet all your mapping needs to develop 30 acres of land into housing subdivisions," where ABC is the proposer's business name.]

12.5.3 TABLE OF CONTENTS

The table of contents should be well formatted with page numbers clearly annotated for each section and subsection of the proposal. The following is a sample of a good table of contents.

TABLE OF CONTENTS
Title Page   1
Table of Contents   2
Executive Summary   3
Bios/Resumes 4
Scope of Work and Approach   6
Schedule and Budget   8
Fee   9
Related Experience   10
Appendix   11

12.5.4 EXECUTIVE SUMMARY

SAMPLE EXECUTIVE SUMMARY
EXECUTIVE SUMMARY
The Objective:
Map land owned by (*Client*) at (*Address*).

- *Need #1:* Permits and licenses
- *Need #2:* Subdivision maps and plans
- *Need #3:* Experienced land developer

The Opportunity:
Map and develop 30 acres of land into housing subdivisions.

- *Goal #1:* Acquire zoning permits and licenses
- *Goal #2:* Carry out land records research
- *Goal #3:* Carry out subdivision survey and mapping
- *Goal #4:* Develop land into housing subdivisions

The Solution:
Hire ABC Surveyors and Land Developers, Inc. to meet all your needs and goals.

- *Recommendation #1:* Allow ABC to prepare zoning permits and license paperwork to submit to ABC County
- *Recommendation #2:* Allow ABC to survey the land
- *Recommendation #3:* Contract with ABC for all phases of land development

12.5.5   Bios/Resumes

Biographical sketches and/or resumes will convey to the client the qualifications and capacity of the firm to carry out the project. The purpose of the section is to highlight the qualifications of the project team (i.e., those to be directly or indirectly in charge of the project). The following sample resumes are adapted from actual student submissions.

SAMPLE BIOSKETCH #1
  Jon Smith, PE, PLS
  CEO and President, ABC Surveyors and Land Developers
  PROJECT ROLE:
  Jon Smith brings more than 20 years' experience to the project. He has diverse background in land matters, having worked on over 200 public and private sector projects. As president, he is responsible for overseeing the company operations. He will be overseer of the operations in this project, ensuring that client expectations are met and quality results are achieved.
  EDUCATION:
  Mr. Smith has a B.S. degree in Geomatics Engineering (2002) from California State University, Fresno.
  PROFESSIONAL LICENSURE:
  Mr. Smith is a licensed Professional Land Surveyor and Principal Engineer in the State of California. PLS license #7776 and PE license # 9998.

SAMPLE BIOSKETCH #2
  Paul Scott, PLS
  Vice President, Party Chief
  PROJECT ROLE:
  Mr. Paul Scott has more than 10 years' experience in ALTA surveys, commercial and residential subdivision, topographic surveys, and boundary surveys, including working with private clients, public entities, state and federal agencies. As Party Chief in this project, he will be in charge of the execution of all field surveys. Notable recent projects of similar experience that he has worked on include the Riverpark Shopping Complex development and Hill Ranch residential development.
  EDUCATION:
  Mr. Scott has a B.S. degree in Geomatics Engineering (2005) from California State University, Fresno.
  PROFESSIONAL LICENSURE:
  Mr. Scott is a licensed Professional Land Surveyor in the State of California. PLS license #9971.

12.5.6   Scope of Work and Approach

This is likely to be the longest section of the proposal and will probably have subsections as well. Discuss in detail each of the items in the RFP and how you intend to tackle them. This is where you underscore project design elements and workflow, with justifications. Use diagrams as appropriate to illustrate your configuration.

Some projects might need only a single workflow diagram with explanation notes. However, most projects will need a systematic explanation of the steps to be applied for each task. Whichever approach is taken, the *scope of work and approach* should aim to do the following:

- Restate project goals and deliverables.
- For each goal, explain how the project objectives will be met.
- Alternatively, explain the overall design and how it will meet all of the project objectives.
- Explain any potential problems or challenges and how you intend to tackle them. What are the alternatives and plans for risk management?

Consider the following template as an example. Numerous other formats can be used to build this section of the proposal.

SCOPE OF WORK AND APPROACH (Sample Template)
TASKS TO BE COMPLETED
Task #1: Acquire Zoning Permits and Licenses
COMPLETION DATE: (*Date and Year*)
RESPONSIBILITY: (*Consultant or Team Members*)
(*Mention why (or if) this will be the first step in the project*)
(*Explain approach to accomplish this task*)
(*Potential problems/challenges and solutions*)
Task #2: Carry out Land Records Research
COMPLETION DATE: (*Date and Year*)
RESPONSIBILITY: (*Consultant or Team Members*)
(*Mention/explain if any prerequisite steps/tasks are required*)
(*Explain approach to accomplish this task*)
(*Potential problems/challenges and solutions*)
Task #3: Carry out Subdivision Survey and Mapping
COMPLETION DATE: (*Date and Year*)
RESPONSIBILITY: (*Consultant or Team Members*)
(*Mention/explain if any prerequisite steps/tasks are required*)
(*Explain approach to accomplish this task*)
(*Potential problems/challenges and solutions*)
Task #4: Develop Land into Housing Subdivisions
COMPLETION DATE: (*Date and Year*)
RESPONSIBILITY: (*Consultant or Team Members*)
(*Mention/explain if any prerequisite steps/tasks are required*)
(*Explain approach to accomplish this task*)
(*Potential problems/challenges and solutions*)

12.5.7 SCHEDULE AND BUDGET

Prepare a project schedule/timeline to indicate when you anticipate starting the project, and how long each task will take. Make a table or Gantt chart of your expected schedule for completing the project (see the example given in Chapter 1, Figure 1.3).

Category	2007	2008	2009	Total
Salaries	720,000	324,000	264,000	1,308,000
Equipment	559,500	128,000	138,000	825.500
Maintenance	119,000	50,000	45,000	214,000
Travel	110,000	40,000	40,000	190,000
Other	30,000	10,000	14,000	54,000
Indirect Costs	53,000	6,000	6,000	65,000
Total:	**1,591,500**	**558,000**	**507,000**	**2,656,500**

FIGURE 12.1　A sample cost summary.

Include a budget summary for the whole duration of the project. The following sample template is for a project spanning more than 1 year (Figure 12.1).

COST SUMMARY (Sample Template)

Cost Category	**Price**
(Insert cost types here)	*(Insert Cost)*
(Insert cost types here)	*(Insert Cost)*
(Insert cost types here)	*(Insert Cost)*
Total Costs:	*(Insert total)*
Ongoing yearly costs	
(Insert cost types here)	*(Insert Cost)*
(Insert cost types here)	*(Insert Cost)*
Total Ongoing	

12.5.8　FEE

Include and explain any fees added to the cost budget. Such fees will include, for example, professional fees, survey records search fees, title company fees, and so forth. This line item can be provided separately or it can be incorporated into the budget and explained.

12.5.9　RELATED EXPERIENCE

List or detail similar past projects that have been completed for other firms and what the results were. Give any useful references or testimonials that the client might use to follow up on your claims. Include testimonials from clients, clippings from newspapers, and so forth.

12.5.10　APPENDIX

Attach additional and third-party material that is too detailed to be included in the main body or a section of the text. The material in the appendix should be clearly annotated and, if needed, mentioned or referenced in the text. The material should be relevant and, most important, helpful to the client. Examples of what things to include in the appendix are (1) explanations or costing of items in the budget; (2) pertinent standard documents; (3) any laws, policy, or local conveyances relevant to the project; (4) legal contract forms or documents; and (5) testimonials from clients about past projects. Don't include an appendix if it is not necessary!

BIBLIOGRAPHY

Morse, L. C. and D. L. Babcock. 2007. *Managing Engineering and Technology.* 4th ed. Upper Saddle River, NJ: Prentice Hall.
PSMJ Resources, Inc. 2006. *Winning Proposals: How to Build Proposals for Extreme Impact.* 3rd ed. Newton, MA: PSMJ Resources.
Stasiowski, F. A. 2003. *Architect's Essentials of Winning Proposals.* 28th ed. New York: Wiley.
White, R. 2010. *Proposal Writing for Government Contracting.* http://www.fed-market.com/articles/government-proposal-writing.shtml (accessed January 15, 2010).

EXERCISES

1. What is Brooks Act, and why was it enacted?
2. What is the difference between an invitation for bid (IFB) and a request for proposal (RFP)?
3. Using a sample RFP, identify:
   a. The client's problem
   b. Your strengths and weaknesses
   c. Your competitors' strengths and weaknesses
      What would be your strategy for a successful proposal?
4. Case Study: Discuss and prepare a proposal in response to the RFP provided in the CD-R. The RFP is titled as follows:
      REQUEST FOR PROPOSALS FOR LAND SURVEYING SERVICES FOR RECORD OF SURVEY OF THE HIGHWAY 111 CORRIDOR WITHIN THE CITY OF PALM DESERT
      The following considerations might be helpful in answering the project question:
   a. Does the above proposal fall under the Brooks Act?
   b. Is the cost budget required with the initial submission?
   c. What kind of unit price schedule is requested (hourly/daily)?
   d. What items will you include in the unit price schedule?
   e. Identify all policies or manuals referred to in this project.
   f. Identify all standards mentioned in the project.
   g. What survey methods or approaches are appropriate for the project based on the provided scope of work?
   h. Are there any environmental issues to deal with?
5. You are the evaluator of a proposal. Discuss your grading in a committee with other evaluators. (You may approach this question by grading a proposal response to a mandatory RFP requirement. Use the RFP in the previous question if needed.)

Part V

Appendices

Appendix A
Resource Guide for Geomatics Projects

All Web site links provided in this appendix have been accessed in the period June 2022–August 2022.

A.1 BOOKS, MANUALS, AND ARTICLES

Applied GPS for Engineers and Project Managers, ISBN 978-0-7844-1150-6, American Society of Civil Engineers, 2011.

ASCE Standard Guidelines for the Collection and Depiction of Existing Subsurface Utility Data, ISBN 978-0-7844-0645-8, American Society of Civil Engineers, 2002.

ASPRS Accuracy Standards for Digital Geospatial Data, Photogrammetric Engineering & Remote Sensing, 1073–1085, 2013.

ASPRS Guidelines for the Procurement of Professional Services, ASPRS, 2009.

ASPRS LIDAR Guidelines: Horizontal Accuracy Reporting, https://www.asprs.org/a/society/committees/standards/Horizontal_Accuracy_Reporting_for_LIDAR_Data.pdf

ASPRS LIDAR Guidelines: Vertical Accuracy Reporting for LIDAR Data, ASPRS, 2004.

ASPRS Interim Standards for Large-Scale Maps, Photogrammetric Engineering & Remote Sensing, 1038–1040, 1989.

ASPRS Manual of Geographic Information Systems, ISBN 1-57083-086-X, 2009.

BLM Guidelines for the Use of Global Navigation Satellite Systems (GNSS) in Cadastral Surveys, 2020.

BLM Standards and Guidelines for Cadastral Surveys Using Global Positioning System Methods, 2001.

FGCS Specifications and Procedures to Incorporate Electronic Digital/Bar-Code Leveling Systems, FGCS, 1995.

Geometric Geodetic Accuracy Standards and Specifications for Using GPS Relative Positioning Techniques, FGCC, 1989.

Geospatial Positioning Accuracy Standards, FGDC, 1998.

GPS Handbook for Professional GPS Users, ISBN 978-90-812754-1-5, 2008.

Land Development Handbook: Planning, Engineering, and Surveying, ISBN 978-0-07-149437-3, New York: McGraw-Hill, 2008.

Manual of Surveying Instructions, U.S. Bureau of Land Management, 2009.

Standards and Specifications for Geodetic Control Networks, FGCC, 1984.

A.2   INTERNET RESOURCES

A.2.1   General Information

drone-laws.com	Drone Laws (by country, state, city)
www.aw-drones.eu	EU Drone Standards Information
www.faa.gov/uas	FAA information on UAS/drones
www.ngs.noaa.gov	GNSS CORS Data and Processing Tools
http://igscb.jpl.nasa.gov	International GNSS Service
www.lsrp.com	Land Surveyor Reference Page
www.suasnews.com	News and information source for unmanned aviation, focusing on small Unmanned Aerial Systems (sUAS)

A.2.2   Data and Software

https://apps.gdgps.net/	NASA Automatic Precise Positioning Service
www.geodesy.noaa.gov/OPUS/	NOAA/NGS/CORS OPUS
www.ngs.noaa.gov/OPUS	NOAA/NGS/CORS OPUS
www.ngs.noaa.gov/TOOLS	NGS Geodetic Toolkit
www.gnss.ga.gov.au/auspos/	Australia's Online GNSS Processing Service
webapp.geod.nrcan.gc.ca/geod/tools-outils/ppp.php	NRCan CSRS-PPP

A.2.3   Standards and Specifications

ascelibrary.org/doi/book/10.1061/9780784406458	ASCE Standard Guidelines for the Collection and Depiction of Existing Subsurface Utility Data
www.asprs.org/committees/standards-committee	ASPRS Standards and Guidelines for Professional Services
www.blm.gov/policy/pim-2020-004	BLM Standards for Positional Accuracy for Cadastral Surveys Conducted Using Global Navigation Satellite Systems (GNSS)
www.aw-drones.eu	EU Drone Standards Information Portal
www.faa.gov/uas	FAA UAS/drone Operating Standards
www.fgdc.gov/standards	Geospatial Positioning Accuracy Standards
www.iclg.com	Mining Laws and Regulations
www.iso.org	International Standards Organization
www.isotc211.org	ISO/TC 211 – Geographic technology standard models & schemas
www.iso.org/committee/54904.html	ISO Technical Committee 211 – Geographic Information/Geomatics
www.fgdc.gov/organization/working-groups-subcommittees/fgcs	FGCC Standards and Specifications for Geodetic Control Networks, GPS Relative Positioning Techniques, and Specifications and Procedures to Incorporate Electronic Digital/Bar-Code Leveling Systems

www.fdacs.gov/content/
download/21300/file/5J-17.052.pdf

www.spatial.nsw.gov.au

5J-17.052 Standards of Practice – Boundary
Survey Requirements (Florida Department
of Agriculture and Consumer Services)
Survey and Drafting Directions for Mining
Surveyors 2020 (NSW Mines)

A.3   GNSS OBSTRUCTION DIAGRAM

National Geodetic Survey Visibility Obstruction Diagram

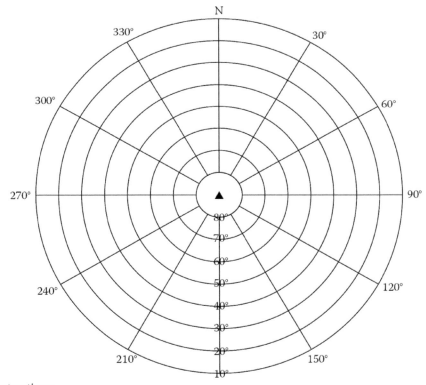

Instructions:

Identify obstructions by azimuth (magnetic) and elevation angle (above horizon) as seen from station mark.
Indicate distance and direction to nearby structures and reflective surfaces (potential multipath sources).

4-char ID: _____ Designation: _____

PID: _____ Location: _____

County: _____ Reconnaissance by: _____

Height above mark, meters: _____ Agency/Company: _____

Phone: (_____) _____ Date: _____

Check if no obstructions above 10 degrees ☐

A.4   GLOBAL SKILLS CHECKLIST

As an engineer in today's global workforce, you must be able to complement your technical skill with many other critical skills, including:

- Being able to analyze other cultures' needs, and design products and services to fit those needs
- Understanding the business environment of the countries where your products and services are made, bought, or sold
- Being aware of customs, laws, and ways of thinking in other countries
- Being self-confident yet humble, listening and learning from people whose value systems differ from yours
- Having some command of the necessary language
- Imagining, forecasting, analyzing, and addressing the potential of local economies and cultures
- Understanding and accepting other cultures' attitudes, behaviors, and beliefs without compromising your own
- Valuing your own cultural heritage while acknowledging its strengths and weaknesses
- Learning about other countries' key business and political leaders and being aware of their philosophies
- Understanding local negotiating strategies
- Understanding e-business and having the electronic skills required for international communication
- Balancing efficient and effective business global travel with family responsibilities
- Understanding international banking and foreign currency exchange
- Being able to unite individuals' diverse skills and interests into a common purpose
- Knowing about other countries' commercial, technical, and cultural developments
- Understanding other locales' environmental issues.

Appendix B
Sample Geodetic Standards

The information in this appendix is reproduced (partially) from the online public domain version at http://www.ngs.noaa.gov/FGCS/tech_pub/.

Author's Note: These standards, dated 1984, are the older standards that existed prior to the advent of Global Positioning System (GPS) surveys. Since 1985, newer standards have been proposed to incorporate the impacts of changing technology such as GPS and Geographic Information System (GIS). There are still many instances where these older standards are applicable in surveying, especially where non-GPS methods are involved.

B.1 INTRODUCTION

(This section is intentionally left blank.)

B.2 STANDARDS

The classification standards of the National Geodetic Control Networks are based on accuracy. This means that when control points in a particular survey are classified, they are certified as having datum values consistent with all other points in the network, not merely those within that particular survey. It is not observation closures within a survey that are used to classify control points, but the ability of that survey to duplicate already established control values. This comparison takes into account models of crustal motion, refraction, and any other systematic effects known to influence the survey measurements.

The National Geographic Survey (NGS) procedure leading to classification covers four steps:

1. The survey measurements, field records, sketches, and other documentation are examined to verify compliance with the specifications for the intended accuracy of the survey. This examination may lead to a modification of the intended accuracy.
2. Results of a minimally constrained least-squares adjustment of the survey measurements are examined to ensure correct weighting of the observations and freedom from blunders.
3. Accuracy measures computed by random error propagation determine the provisional accuracy. If the provisional accuracy is substantially different from the intended accuracy of the survey, then the provisional accuracy supersedes the intended accuracy.

4. A variance factor ratio for the new survey combined with network data is computed by the iterated almost unbiased estimator (IAUE) method. If the variance factor ratio is reasonably close to 1.0 (typically less than 1.5), then the survey is considered to check with the network, and the survey is classified with the provisional (or intended) accuracy. If the variance factor ratio is much greater than 1.0 (typically 1.5 or greater), then the survey is considered to not check with the network, and both the survey and network measurements will be scrutinized for the source of the problem.

B.2.1 HORIZONTAL CONTROL NETWORK STANDARDS

When a horizontal control point is classified with a particular order and class, NGS certifies that the geodetic latitude and longitude of that control point bear a relation of specific accuracy to the coordinates of all other points in the horizontal control network. This relation is expressed as a distance accuracy, 1:a. A distance accuracy is the ratio of the relative positional error of a pair of control points to the horizontal separation of those points.

A distance accuracy, 1:a, is computed from a minimally constrained, correctly weighted, least-squares adjustment by:

$$a = d/s,$$

where

a = distance accuracy denominator
s = propagated standard deviation of distance between survey points obtained from the least-squares adjustment
d = distance between survey points

The distance accuracy pertains to all pairs of points (but in practice is computed for a sampling of pairs of points). The worst distance accuracy (smallest denominator) is taken as the provisional accuracy. If this is substantially larger or smaller than the intended accuracy, then the provisional accuracy takes precedence.

TABLE B.1
Distance Accuracy Standards

Classification	Minimum Distance Accuracy
First-order	1:100,000
Second-order, class I	1:50,000
Second-order, class II	1:20,000
Third-order, class I	1:10,000
Third-order, class II	1:5,000

As a test for systematic errors, the variance factor ratio of the new survey is computed by the IAUE method. This computation combines the new survey measurements with existing network data, which are assumed to be correctly weighted and free of systematic error. If the variance factor ratio is substantially greater than unity when the survey does not check with the network, and both the survey and the network data will be examined by NGS.

Computer simulations performed by NGS have shown that a variance factor ratio greater than 1.5 typically indicates systematic errors between the survey and the network. Setting a cutoff value higher than this could allow undetected systematic error to propagate into the national network. On the other hand, a higher cutoff value might be considered if the survey has only a small number of connections to the network, because this circumstance would tend to increase the variance factor ratio.

In some situations, a survey has been designed in which different sections provide different orders of control. For these multiorder surveys, the computed distance accuracy denominators should be grouped into sets appropriate to the different parts of the survey. Then, the smallest value of a in each set is used to classify the control points of that portion. If there are sufficient connections to the network, several variance factor ratios, one for each section of the survey, should be computed.

B.2.1.1 Horizontal Example

Suppose a survey with an intended accuracy of first-order (1:100,000) has been performed. A series of propagated distance accuracies from a minimally constrained adjustment is now computed.

Line	s (m)	d (m)	1:a
1–2	0.141	17,107	1:121,326
1–3	0.170	20,123	1:118,371
2–3	0.164	15,505	1:94,543
....	.	.	.
....	.	.	.
....	.	.	.

Suppose that the worst distance accuracy is 1:94,543. This is not substantially different from the intended accuracy of 1:100,000, which would therefore have precedence for classification. It is not feasible to precisely quantify "substantially different." Judgment and experience are determining factors.

Now assume that a solution combining survey and network data has been obtained, and that a variance factor ratio of 1.2 was computed for the survey. This would be reasonably close to unity and would indicate that the survey checks with the network. The survey would then be classified as first-order using the intended accuracy of 1:100,000.

However, if a variance factor of, say, 1.9 was computed, the survey would not check with the network. Both the survey and network measurements then would have to be scrutinized to find the problem.

B.2.1.2   Monumentation

Control points should be part of the National Geodetic Horizontal Network only if they possess permanence, horizontal stability with respect to the Earth's crust, and a horizontal location that can be defined as a point. A 30-cm-long wooden stake driven into the ground, for example, would lack both permanence and horizontal stability. A mountain peak is difficult to define as a point. Typically, corrosion-resistant metal disks set in a large concrete mass have the necessary qualities. First-order and second-order, class I, control points should have an underground mark, at least two monumented reference marks at right angles to one another, and at least one monumented azimuth mark no less than 400 m from the control point. Replacement of a temporary mark by a more permanent mark is not acceptable unless the two marks are connected in timely fashion by survey observations of sufficient accuracy. Detailed information may be found in *C&GS Special Publication 247*, "Manual of Geodetic Triangulation."

B.2.2   VERTICAL CONTROL NETWORK STANDARDS

When a vertical control point is classified with a particular order and class, NGS certifies that the orthometric elevation at that point bears a relation of specific accuracy to the elevations of all other points in the vertical control network. That relation is expressed as an elevation difference accuracy, b. An elevation difference accuracy is the relative elevation error between a pair of control points that is scaled by the square root of their horizontal separation traced along existing level routes (Table B.2).

An elevation difference accuracy, b, is computed from a minimally constrained, correctly weighted, least-squares adjustment by

$$b = S/\sqrt{d}.$$

where

d = approximate horizontal distance in kilometers between control point positions traced along existing level routes

S = propagated standard deviation of elevation difference in millimeters between survey control points obtained from the least-squares adjustment

Note that the units of b are $(\text{mm})/\sqrt{(km)}$.

TABLE B.2
Elevation Accuracy Standards

Classification	Maximum Elevation Difference Accuracy
First-order, class I	0.5
First-order, class II	0.7
Second-order, class I	1.0
Second-order, class II	1.3
Third-order	2.0

The elevation difference accuracy pertains to all pairs of points (but in practice is computed for a sample). The worst elevation difference accuracy (largest value) is taken as the provisional accuracy. If this is substantially larger or smaller than the intended accuracy, then the provisional accuracy takes precedence.

As a test for systematic errors, the variance factor ratio of the new survey is computed by the IAUE method. This computation combines the new survey measurements with existing network data, which are assumed to be correctly weighted and free of systematic error. If the variance factor ratio is substantially greater than unity, then the survey does not check with the network, and both the survey and the network data will be examined by NGS.

Computer simulations performed by NGS have shown that a variance factor ratio greater than 1.5 typically indicates systematic errors between the survey and the network. Setting a cutoff value higher than this could allow undetected systematic error to propagate into the national network. On the other hand, a higher cutoff value might be considered if the survey has only a small number of connections to the network, because this circumstance would tend to increase the variance factor ratio.

In some situations, a survey has been designed in which different sections provide different orders of control. For these multiorder surveys, the computed elevation difference accuracies should be grouped into sets appropriate to the different parts of the survey. Then, the largest value of b in each set is used to classify the control points of that portion, as discussed earlier. If there are sufficient connections to the network, several variance factor ratios, one for each section of the survey, should be computed.

B.2.2.1   Vertical Example

Suppose a survey with an intended accuracy of second-order, class II has been performed. A series of propagated elevation difference accuracies from a minimally constrained adjustment is now computed.

Line	S (mm)	d (km)	b (mm)/$\sqrt{(km)}$
1–2	1.574	1.718	1.20
1–3	1.743	2.321	1.14
2–3	2.647	4.039	1.32
....	.	.	.
....	.	.	.
....	.	.	.

Suppose that the worst elevation difference accuracy is 1.32. This is not substantially different from the intended accuracy of 1.3, which would therefore have precedence for classification. It is not feasible to precisely quantify "substantially different." Judgment and experience are determining factors.

Now assume that a solution combining survey and network data has been obtained, and that a variance factor ratio of 1.2 was computed for the survey. This would be reasonably close to unity and would indicate that the survey checks with the network.

The survey would then be classified as second-order, class II, using the intended accuracy of 1.3.

However, if a survey variance factor ratio of, say, 1.9 was computed, the survey would not check with the network. Both the survey and network measurements then would have to be scrutinized to find the problem.

B.2.2.2 Monumentation

Control points should be part of the National Geodetic Vertical Network only if they possess permanence, vertical stability with respect to the Earth's crust, and a vertical location that can be defined as a point. A 30-cm-long wooden stake driven into the ground, for example, would lack both permanence and vertical stability. A rooftop lacks stability and is difficult to define as a point. Typically, corrosion-resistant metal disks set in large rock outcrops or long metal rods driven deep into the ground have the necessary qualities. Replacement of a temporary mark by a more permanent mark is not acceptable unless the two marks are connected in timely fashion by survey observations of sufficient accuracy. Detailed information may be found in *NOAA Manual NOS NGS* 1, "Geodetic Bench Marks."

B.2.3 Gravity Control Network Standards

When a gravity control point is classified with a particular order and class, NGS certifies that the gravity value at that control point possesses a specific accuracy.

Gravity is commonly expressed in units of milligals (mGal) or microgals (μGal) equal, respectively, to (10^{-5}) m/s^2, and (10^{-8}) m/s^2. Classification order refers to measurement accuracies and class to site stability.

When a survey establishes only new points, and where only absolute measurements are observed, then each survey point is classified independently. The standard deviation from the mean of measurements observed at that point is corrected by the error budget for noise sources in accordance with the following formula:

$$c^2 = \sum_{i+1}^{n} \frac{(x_i - x_m)^2}{n-1} + e^2$$

TABLE B.3
Gravity Accuracy Standards

Classification	Gravity Accuracy (μGal)
First-order, class I	20 (subject to stability verification)
First-order, class II	20
Second-order	50
Third-order	100

---

where

c = gravity accuracy
x_i = gravity measurement
n = number of measurements

$$x_m = \left(\sum_{i=1}^{n} x_i \right) / (n)$$

e = external random error

The value obtained for c is then compared directly against the gravity accuracy standards table.

When a survey establishes points at which both absolute and relative measurements are made, the absolute determination ordinarily takes precedence and the point is classified accordingly. (However, see Example D later for an exception.)

When a survey establishes points where only relative measurements are observed, and where the survey is tied to the National Geodetic Gravity Network, then the gravity accuracy is identified with the propagated gravity standard deviation from a minimally constrained, correctly weighted, least-squares adjustment.

The worst gravity accuracy of all the points in the survey is taken as the provisional accuracy. If the provisional accuracy exceeds the gravity accuracy limit set for the intended survey classification, then the survey is classified using the provisional accuracy.

As a test for systematic errors, the variance factor ratio of the new survey is computed by the IAUE method. This computation combines the new survey measurements with existing network data, which are assumed to be correctly weighted and free of systematic error. If the variance factor ratio is substantially greater than unity, then the survey does not check with the network, and both the survey and the network data will be examined by NGS.

Computer simulations performed by NGS have shown that a variance factor ratio greater than 1.5 typically indicates systematic errors between the survey and the network. Setting a cutoff value higher than this could allow undetected systematic error to propagate into the national network. On the other hand, a higher cutoff value might be considered if the survey has only a minimal number of connections to the network, because this circumstance would tend to increase the variance factor ratio.

In some situations, a survey has been designed in which different sections provide different orders of control. For these multiorder surveys, the computed gravity accuracies should be grouped into sets appropriate to the different parts of the survey. Then, the largest value of c in each set is used to classify the control points of that portion, as discussed earlier. If there are sufficient connections to the network, several variance factor ratios, one for each part of the survey, should be computed.

B.2.3.1   Gravity Examples

Example A. Suppose a gravity survey using absolute measurement techniques has been performed. These points are then unrelated. Consider one of these survey points.

$$\text{Assume } n = 750$$

$$\sum_{i=1}^{750} (x_i - x_m)^2 = .169 \text{ mGal}^2$$

$$e = 5 \text{ mGal}$$

$$c^2 = \frac{0.169}{750 - 1} + (.005)^2$$

$$c = 16 \text{ mGal}$$

The point is then classified as first-order, class II.

Example B. Suppose a relative gravity survey with an intended accuracy of second-order (50 µGal) has been performed. A series of propagated gravity accuracies from a minimally constrained adjustment is now computed.

Standard	Gravity Station (Deviation [µGal])
1	38
2	44
3	55
.	.
.	.
.	.

Suppose that the worst gravity accuracy was 55 µGal. This is worse than the intended accuracy of 50 µGal. Therefore, the provisional accuracy of 55 µGal would have precedence for classification, which would be set to third-order.

Now assume that a solution combining survey and network data has been obtained and that a variance factor of 1.2 was computed for the survey. This would be reasonably close to unity and would indicate that the survey checks with the network. The survey would then be classified as third-order using the provisional accuracy of 55 µGal.

However, if a variance factor of, say, 1.9 was computed, the survey would not check with the network. Both the survey and network measurements then would have to be scrutinized to find the problem.

Example C. Suppose a survey consisting of both absolute and relative measurements has been made at the same points. Assume the absolute observation at one of the points yielded a classification of first-order, class II, whereas the relative measurements produced a value to second-order standards. The point in question would be classified as first-order, class II, in accordance with the absolute observation.

Example D. Suppose we have a survey similar to Example C, where the absolute measurements at a particular point yielded a third-order classification due to an unusually noisy observation session, but the relative measurements still satisfied the second-order standard. The point in question would be classified as second-order, in accordance with the relative measurements.

B.2.3.2 Monumentation

Control points should be part of the National Geodetic Gravity Network only if they possess permanence, horizontal and vertical stability with respect to the Earth's crust, and a horizontal and vertical location that can be defined as a point. For all orders of accuracy, the mark should be imbedded in a stable platform such as flat, horizontal concrete. For first-order, class I stations, the platform should be imbedded in stable, hard rock, and checked at least twice for the first year to ensure stability. For first-order, class II stations, the platform should be located in an extremely stable environment such as the concrete floor of a mature structure. For second- and third-order stations, standard benchmark monumentation is adequate. Replacement of a temporary mark by a more permanent mark is not acceptable unless the two marks are connected in timely fashion by survey observations of sufficient accuracy. Detailed information is given in *NOAA Manual NOS NGS* 1, "Geodetic Bench Marks." Monuments should not be near sources of electromagnetic interference.

It is recommended, but not necessary, to monument third-order stations. However, the location associated with the gravity value should be recoverable, based upon the station description.

B.3   SPECIFICATIONS

B.3.1   Introduction

All measurement systems regardless of their nature have certain common qualities. Because of this, the measurement system specifications follow a prescribed structure as outlined next. These specifications describe the important components and state permissible tolerances used in a general context of accurate surveying methods. The user is cautioned that these specifications are not substitutes for manuals that detail recommended field operations and procedures.

The observations will have spatial or temporal relationships with one another as given in the "Network Geometry" section. In addition, this section specifies the frequency of incorporation of old control into the survey. Computer simulations could be performed instead of following the "Network Geometry" and "Field Procedures" specifications. However, the user should consult the National Geodetic Survey before undertaking such a departure from the specifications.

The "Instrumentation" section describes the types and characteristics of the instruments used to make observations. An instrument must be able to attain the precision requirements given in "Field Procedures."

The section "Calibration Procedures" specifies the nature and frequency of instrument calibration. An instrument must be calibrated whenever it has been damaged or repaired.

The "Field Procedures" section specifies particular rules and limits to be met while following an appropriate method of observation. For a detailed account of how to perform observations, the user should consult the appropriate manuals.

Because NGS will perform the computations described under "Office Procedures," it is not necessary for the user to do them. However, these computations provide valuable checks on the survey measurements that could indicate the need for some reobservations. This section specifies commonly applied corrections to observations, and computations that monitor the precision and accuracy of the survey. It also discusses the correctly weighted, minimally constrained least-squares adjustment used to ensure that the survey work is free from blunders and able to achieve the intended accuracy. Results of the least-squares adjustment are used in the quality control and accuracy classification procedures. The adjustment performed by NGS will use models of error sources, such as crustal motion, when they are judged to be significant to the level of accuracy of the survey.

B.3.2 TRIANGULATION

Triangulation is a measurement system comprised of joined or overlapping triangles of angular observations supported by occasional distance and astronomic observations. Triangulation is used to extend horizontal control.

B.3.2.1   Network Geometry

Order	First	Second	Second	Third	Third
Class		I	II	I	II
Station spacing not less than (km)	15	10	5	0.5	0.5
Average minimum distance angle[a] of figures not less than	40°	35°	30°	30°	25°
Minimum distance angle[a] of all figures not less than	30°	25°	25°	20°	20°
Base line spacing not more than (triangles)	5	10	12	15	15
Astronomic azimuth spacing not more than (triangles)	8	10	10	12	15

[a]   Distance angle is angle opposite the side through which distance is propagated.

The new survey is required to tie to at least four network control points spaced well apart. These network points must have datum values equivalent to or better than the intended order (and class) of the new survey. For example, in an arc of triangulation, at least two network control points should be occupied at each end of the arc. Whenever the distance between two new unconnected survey points is less than 20% of the distance between those points traced along existing or new connections, then a direct connection should be made between those two survey points. In addition, the survey should tie into any sufficiently accurate network control points within the station spacing distance of the survey. These network stations should be occupied and sufficient observations taken to make these stations integral parts of the survey. Nonredundant geodetic connections to the network stations are not considered sufficient ties. Nonredundantly determined stations are not allowed. Control stations should not be determined by intersection or resection methods. Simultaneous reciprocal vertical angles or geodetic leveling are observed along baselines. A base line need not be observed if other baselines of sufficient accuracy were observed within the base line spacing specification in the network, and similarly for astronomic azimuths.

B.3.2.2   Instrumentation

Only properly maintained theodolites are adequate for observing directions and azimuths for triangulation. Only precisely marked targets, mounted stably on tripods or supported towers, should be employed. The target should have a clearly defined center, resolvable at the minimum control spacing. Optical plummets or collimators are required to ensure that the theodolites and targets are centered over the marks. Microwave-type electronic distance measurement (EDM) equipment is not sufficiently accurate for measuring higher-order baselines.

Order	First	Second	Second	Third	Third
Class		I	II	I	II
Theodolite, least count	0.2″	0.2″	1.0″	1.0″	1.0″

B.3.2.3   Calibration Procedures

Each year and whenever the difference between direct and reverse readings of the theodolite departs from 180° by more than 30″, the instrument should be adjusted for collimation error. Readjustment of the crosshairs and the level bubble should be done whenever their misadjustments affect the instrument reading by the amount of the least count.

All EDM devices and retroreflectors should be serviced regularly and checked frequently over lines of known distances. The National Geodetic Survey has established specific calibration baselines for this purpose. EDM instruments should be calibrated annually and frequency checks made semiannually.

B.3.2.4   Field Procedures

Theodolite observations for first-order and second-order, class I surveys may only be made at night. Reciprocal vertical angles should be observed at times of best atmospheric conditions (between noon and late afternoon) for all orders of accuracy. EDMs need a record at both ends of the line of wet and dry bulb temperatures to ±1°C, and barometric pressure to ±5 mm of mercury. The theodolite and targets should be centered to within 1 mm over the survey mark or eccentric point.

Measurements of astronomic latitude and longitude are not required in the United States, except perhaps for first-order work, because sufficient information for determining deflections of the vertical exists. Detailed procedures can be found in Hoskinson and Duerksen (1952).

B.3.2.5   Office Procedures

A minimally constrained least-squares adjustment will be checked for blunders by examining the normalized residuals. The observation weights will be checked by inspecting the postadjustment estimate of the variance of unit weight. Distance standard errors computed by error propagation in this correctly weighted least-squares adjustment will indicate the provisional accuracy classification. A survey variance

Order	First	Second	Second	Third	Third
Class		I	II	I	II
Directions					
Number of positions	16	16	8 or 12[a]	4	2
Standard deviation of mean not to exceed	0.4″	0.5″	0.8″	1.2″	2.0″
Rejection limit from the mean	4″	4″	5″	5″	5″
Reciprocal Vertical Angles (Along Distance Sight Path)					
Number of independent observations direct/reverse	3	3	2	2	2
Maximum spread	10″	10″	10″	10″	20″
Maximum time interval between reciprocal angles (hr)	1	1	1	1	1
Astronomic Azimuths					
Observations per night	16	16	16	8	4
Number of nights	2	2	1	1	1
Standard deviation of mean not to exceed	0.45″	0.45″	0.6″	1.0″	1.7″
Rejection limit from the mean	5″	5″	5″	6″	6″
Electro-Optical Distances					
Minimum number of days	2[b]	2[b]	1	1	1
Minimum number of measurements/day	2[c]	2[c]	2[c]	1	1
Minimum number of concentric observations/measurement	2	2	1	1	1
Minimum number of offset observations/measurement	2	2	2	1	1
Maximum difference from mean of observations/(mm)	40	40	50	60	60
Minimum number of readings/observation (or equivalent)	10	10	10	10	10
Maximum difference from mean of readings (mm)	d	d	d	d	d
Infrared Distances					
Minimum number of days	—	2[b]	1	1	1
Minimum number of measurements	—	2[c]	2[c]	1	1
Minimum number of concentric observations/measurement	—	1	1	1	1
Minimum number of offset observations/measurement	—	2	1	1	1
Maximum difference from mean of observations/(mm)	—	5	5	10	10
Minimum number of readings/observation (or equivalent)	—	10	10	10	10
Maximum difference from mean of readings (mm)	—	d	d	d	d
Microwave Distances					
Minimum number of measurements	—	—	—	2	1
Minimum time span between measurements (hr)	—	—	—	8	—
Maximum difference between measurements (mm)	—	—	—	100	—
Minimum number of concentric observations/ measurement	—	—	—	2[e]	1[e]
Maximum difference from mean of observations/(mm)	—	—	—	100	150
Minimum number of readings/observation (or equivalent)	—	—	—	20	20
Maximum difference from mean of readings (mm)	—	—	—	a	a

a 8 if 0.2″, 12 if 1.0″ resolution.
b Two or more instruments.
c One measurement at each end of the line.
d As specified by manufacturer.
e Carried out at both ends of the line.

Order	First	Second	Second	Third	Third
Class		I	II	I	II
Triangle Closure					
Average not to exceed	1.0″	1.2″	2.0″	3.0″	5.0″
Maximum not to exceed	3″	3″	5″	5″	10″
Side Checks					
Mean absolute correction by side equation not to exceed	0.3″	0.4″	0.6″	0.8″	2.0″

factor ratio will be computed to check for systematic error. The least-squares adjustment will use models which account for the following:

Semimajor axis of the ellipsoid	(a = 6378137 m)
Reciprocal flattening of the ellipsoid	(1/f = 298.257222)
Mark elevation above mean sea level	(known to ±1 m)
Geoid height	(known to ±6 m)
Deflection of the vertical	(known to ±3″)
Geodesic correction	
Skew normal correction	
Height of instrument	
Height of target	
Sea level correction	
Arc correction	
Geoid height correction	
Second velocity correction	
Crustal motion.	

B.3.3 TRAVERSE

Traverse is a measurement system comprised of joined distance and theodolite observations supported by occasional astronomic observations. Traverse is used to densify horizontal control.

B.3.3.1 Network Geometry

The new survey is required to tie to a minimum number of network control points spaced well apart. These network points must have datum values equivalent to or better than the intended order (and class) of the new survey. Whenever the distance between two new unconnected survey points is less than 20% of the distance between those points traced along existing or new connections, then a direct connection must be made between those two survey points. In addition, the survey should tie into any sufficiently accurate network control points within the station spacing distance of the survey. These ties must include EDM or taped distances. Nonredundant geodetic connections to the network stations are not considered sufficient ties. Nonredundantly

Order	First	Second	Second	Third	Third
Class		I	II	I	II
Station spacing not less than (km)	10	4	2	0.5	0.5
Maximum deviation of main traverse from straight line	20°	20°	25°	30°	40°
Minimum number of benchmark ties	2	2	2	2	2
Benchmark tie spacing not more than (segments)	6	8	10	15	20
Astronomic azimuth spacing not more than (segments)	6	12	20	25	40
Minimum number of network control points	4	3	2	2	2

determined stations are not allowed. Reciprocal vertical angles or geodetic leveling are observed along all traverse lines.

B.3.3.2 Instrumentation

Only properly maintained theodolites are adequate for observing directions and azimuths for traverse. Only precisely marked targets, mounted stably on tripods or supported towers, should be employed. The target should have a clearly defined center, resolvable at the minimum control spacing. Optical plummets or collimators are required to ensure that the theodolites and targets are centered over the marks. Microwave-type EDM equipment is not sufficiently accurate for measuring first-order traverses.

Order	First	Second	Second	Third	Third
Class		I	II	I	II
Theodolite, least count	0.2″	1.0″	1.0″	1.0″	1.0″

B.3.3.3 Calibration Procedures

Each year and whenever the difference between direct and reverse readings of the theodolite departs from 180° by more than 30″, the instrument should be adjusted for collimation error. Readjustment of the crosshairs and the level bubble should be done whenever their misadjustments affect the instrument reading by the amount of the least count.

All electronic distance-measuring devices and retroreflectors should be serviced regularly and checked frequently over lines of known distances. The National Geodetic Survey has established specific calibration baselines for this purpose. EDM instruments should be calibrated annually and frequency checks made semiannually.

B.3.3.4 Field Procedures

Theodolite observations for first-order and second-order, class I surveys may be made only at night. EDMs need a record at both ends of the line of wet and dry bulb temperatures to ±1°C, and barometric pressure to ±5 mm of mercury. The theodolite, EDM, and targets should be centered to within 1 mm over the survey mark or eccentric point.

Order	First	Second	Second	Third	Third
Class		**I**	**II**	**I**	**II**
Directions					
Number of positions	16	8 or 12[a]	6 or 8[b]	4	2
Standard deviation of mean not to exceed	0.4″	0.5″	0.8″	1.2″	2.0″
Rejection limit from the mean	4″	5″	5″	5″	5″
Reciprocal Vertical Angles (Along Distance Sight Path)					
Number of independent observations, direct/reverse	3	3	2	2	2
Maximum spread	10″	10″	10″	10″	20″
Maximum time interval between reciprocal angles (hr)	1	1	1	1	1
Astronomic Azimuths					
Observations per night	16	16	12	8	4
Number of nights	2	2	1	1	1
Standard deviation of mean not to exceed	0.45″	0.45″	0.6″	1.0″	1.7″
Rejection limit from the mean	5″	5″	5″	6″	6″
Electro-Optical Distances					
Minimum number of measurements	1	1	1	1	1
Minimum number of concentric observations/ measurement	1	1	1	1	1
Minimum number of offset observations/measurement	1	1	—	—	—
Maximum difference from mean of observations/(mm)	60	60	—	—	—
Minimum number of readings/observation (or equivalent)	10	10	10	10	10
Maximum difference from mean of readings (mm)	c	c	c	c	c
Infrared Distances					
Minimum number of measurements	1	1	1	1	1
Minimum number of concentric observations/ measurement	1	1	1	1	1
Minimum number of offset observations/measurement	1	1	1d	—	—
Maximum difference from mean of observations (mm)	10	10	10d	—	—
Minimum number of readings/observation (or equivalent)	10	10	10	10	10
Maximum difference from mean of readings (mm)	c	c	c	c	c
Microwave Distances					
Minimum number of measurements	—	1	1	1	1
Minimum number of concentric observations/ measurement	—	2e	1e	1e	1e
Maximum difference from mean of observations/(mm)	—	150	150	200	200
Minimum number of readings/observation (or equivalent)	—	20	20	10	10
Maximum difference from mean of readings (mm)	—	c	c	c	c

[a] *8 if 0.2″, 12 if 1.0″ resolution.*
[b] *6 if 0.2″, 8 if 1.0″ resolution.*
[c] *As specified by manufacturer.*
[d] *Only if decimal reading near 0 or high 9's.*
[e] *Carried out at both ends of the line.*

B.3.3.5 Office Procedures

Order	First	Second	Second	Third	Third
Class		I	II	I	II
Azimuth closure at azimuth check point (seconds of arc)	$1.7\sqrt{N}$	$3.0\sqrt{N}$	$4.5\sqrt{N}$	$10.0\sqrt{N}$	$12.0\sqrt{N}$
Position closure after azimuth adjustment[a]	$0.04\sqrt{K}$ or 1;100,000	$0.08\sqrt{K}$ or 1:50,000	$0.20\sqrt{K}$ or 1:20,000	$0.40\sqrt{K}$ or 1:10,000	$0.80\sqrt{K}$ or 1:5,000

Note: N is the number of segments, and K is route distance in km.

[a] *The expression containing the square root is designed for longer lines where higher proportional accuracy is required. Use the formula that gives the smallest permissible closure. The closure (e.g., 1:100,000) is obtained by computing the difference between the computed and fixed values, and dividing this difference by K. Note: Do not confuse closure with distance accuracy of the survey.*

A minimally constrained least-squares adjustment will be checked for blunders by examining the normalized residuals. The observation weights will be checked by inspecting the postadjustment estimate of the variance of unit weight. Distance standard errors computed by error propagation in this correctly weighted least-squares adjustment will indicate the provisional accuracy classification. A survey variance factor ratio will be computed to check for systematic error. The least-squares adjustment will use models which account for the following:

Semimajor axis of the ellipsoid	(a = 6,378,137 m)
Reciprocal flattening of the ellipsoid	(1/f = 298.257222)
Mark elevation above mean sea level	(known to ±1 m)
Geoid height	(known to ±6 m)
Deflection of the vertical	(known to ±3″)
Geodesic correction	
Skew normal correction	
Height of instrument	
Height of target	
Geodesic correction	
Sea level correction	
Arc correction	
Geoid height correction	
Second velocity correction	
Crustal motion.	

B.3.4 GEODETIC LEVELING

Geodetic leveling is a measurement system comprised of elevation differences observed between nearby rods. Leveling is used to extend vertical control.

B.3.4.1   Network Geometry

Order	First	Second	Second	Third	Third
Class		I	II	I	II
Benchmark spacing not more than (km)	3	3	3	3	3
Average benchmark spacing not more than (km)	1.6	1.6	1.6	3.0	3.0
Line length between network control points not more than (km)	300	100	50	50	25
					(double run)
				25	10
					(single run)

New surveys are required to tie to existing network benchmarks at the beginning and end of the leveling line. These network benchmarks must have an order (and class) equivalent to or better than the intended order (and class) of the new survey. First-order surveys are required to perform check connections to a minimum of six benchmarks, three at each end. All other surveys require a minimum of four check connections, two at each end. "Check connection" means that the observed elevation difference agrees with the adjusted elevation difference within the tolerance limit of the new survey. Checking the elevation difference between two benchmarks located on the same structure, or so close together that both may have been affected by the same localized disturbance, is not considered a proper check. In addition, the survey is required to connect to any network control points within 3 km of its path. However, if the survey is run parallel to existing control, then the following table specifies the maximum spacing of extra connections between the survey and the control. At least one extra connection should always be made.

Distance, Survey to Network	Maximum Spacing of Extra Connections (km)
0.5 km or less	5
0.5 km to 2.0 km	10
2.0 km to 3.0 km	20

B.3.4.2   Instrumentation

Only a compensator or tilting leveling instrument with an optical micrometer should be used for first-order leveling. Leveling rods should be one piece. Wooden or metal rods may be employed only for third-order work. A turning point consisting of a steel turning pin with a driving cap should be utilized. If a steel pin cannot be driven, then a turning plate ("turtle") weighing at least 7 kg should be substituted. In situations allowing neither turning pins nor turning plates (sandy or marshy soils), a long wooden stake with a double-headed nail should be driven to a firm depth.

Order	First	Second	Second	Third	Third
Class		I	II	I	II
Leveling Instrument					
Minimum repeatability of line of sight	0.25″	0.25″	0.50″	0.50″	1.00″
Leveling rod construction	IDS	IDS	IDS[a] or ISS	ISS	Wood or Metal
Instrument and Rod Resolution (Combined)					
Least count (mm)	0.1	0.1	0.5–1.0[b]	1.0	1.0

a If optical micrometer is used.
b 1.0 mm if 3-wire method, 0.5 mm if optical micrometer.
IDS, invar, double scale; ISS, invar, single scale.

B.3.4.3 Calibration Procedures

Order	First	Second	Second	Third	Third
Class		I	II	I	II
Leveling Instrument					
Maximum collimation error, single line of sight (mm/m)	0.05	0.05	0.05	0.05	0.10
Maximum collimation error, reversible compensator-type instruments, mean of two lines of sight (mm/m)	0.02	0.02	0.02	0.02	0.04
Time interval between collimation error determination not longer than (days)					
Reversible compensator	7	7	7	7	7
Other types	1	1	1	1	7
Maximum angular difference between two lines of sight, reversible compensator	40″	40″	40″	40″	60″
Leveling Rod					
Maximum scale calibration standard	N	N	N	M	M
Time interval between scale calibration (yr)	1	1	—	—	—
Leveling rod bubble verticality maintained to within	10′	10′	10′	10′	10′

N, national standard; M, manufacturer's standard.

Compensator-type instruments should be checked for proper operation at least every 2 weeks of use. Rod calibration should be repeated whenever the rod is dropped or damaged in any way. Rod levels should be checked for proper alignment once a week. The manufacturer's calibration standard should, as a minimum, describe scale behavior with respect to temperature.

B.3.4.4 Field Procedures

Order	First	Second	Second	Third	Third
Class		I	II	I	II
Minimum observation method	Micrometer	Micrometer	Micrometer or 3-wire	3-wire	Center wire
Section naming	SRDS or DR or SP	SRDS or DR or SP	SRDS or DR[a] or SP	SRDS or DR[b]	SRDS or DR[c]
Difference of forward and backward sight lengths never to exceed					
per setup (m)	2	5	5	10	10
per section (m)	4	10	10	10	10
Maximum sight length (m)	50	60	60	70	90
Minimum ground clearance of line of sight (m)	0.5	0.5	0.5	0.5	0.5
Even number of setups when not using leveling rods with detailed calibration	Yes	Yes	Yes	Yes	Yes
Determine temperature gradient for the vertical range of the line of sight at each setup	Yes	Yes	Yes	—	—
Maximum section miclosure (mm)	$3\sqrt{D}$	$4\sqrt{D}$	$6\sqrt{D}$	$8\sqrt{D}$	$12\sqrt{D}$
Maximum loop miclosure (mm)	$4\sqrt{E}$	$5\sqrt{E}$	$6\sqrt{E}$	$8\sqrt{E}$	$12\sqrt{E}$
Single-Run Methods					
Reverse direction of single runs every half day	Yes	Yes	Yes	—	—
Nonreversible Compensator Leveling Instruments					
Off-level/relevel instrument between observing the high and low rod scales	Yes	Yes	Yes	—	—
3-Wire Method					
Reading check (difference between top and bottom intervals) for one setup not to exceed (tenths of rod units)	—	—	2	2	3
Read rod 1 first in alternate setup method	—	—	Yes	Yes	Yes
Double-Scale Rods					
Low-high scale elevations difference for one setup not to exceed (mm)					
With reversible compensator	0.40	1.00	1.00	2.00	2.00
Other instrument types:					
Half-centimeter rods	0.25	0.30	0.60	0.70	1.30
Full-centimeter rods	0.30	0.30	0.60	0.70	1.30

[a] *Must double run when using 3-wire method.*
[b] *May single-run if line length between network control points is less than 25 km.*
[c] *May single-run if line length between network control points is less than 25 km.*
SRDS, single-run, double simultaneous procedure; DR, double run; SP, SPur, less than 25 km, double run; D, shortest length of section (one-way) in km; E, perimeter of loop in km.

Double-run leveling may always be used, but single-run leveling done with the double simultaneous procedure may be used only where it can be evaluated by loop closures. Rods should be leap-frogged between setups (alternate setup method). The date, beginning and ending times, cloud coverage, air temperature (to the nearest degree), temperature scale, and average wind speed should be recorded for each section plus any changes in the date, instrumentation, observer, or time zone. The instrument need not be off-leveled/releveled between observing the high and low scales when using an instrument with a reversible compensator. The low-high scale difference tolerance for a reversible compensator is used only for the control of blunders.

With double-scale rods, the following observing sequence should be used:

Backsight, low-scale
Backsight, stadia
Foresight, low-scale
Foresight, stadia
Off-level/relevel or reverse compensator
Foresight, high scale
Backsight, high scale.

B.3.4.5 Office Procedures

Order	First	Second	Second	Third	Third
Class		I	II	I	II
Section Misclosures Backward and Forward					
Algebraic sum of all corrected section misclosures of a leveling line not to exceed (mm)	$3\sqrt{D}$	$4\sqrt{D}$	$6\sqrt{D}$	$8\sqrt{D}$	$12\sqrt{D}$
Section misclosure not to exceed (mm)	$3\sqrt{E}$	$4\sqrt{E}$	$6\sqrt{E}$	$8\sqrt{E}$	$12\sqrt{E}$
Loop Misclosures					
Algebraic sum of all corrected misclosures not to exceed (mm)	$4\sqrt{F}$	$5\sqrt{F}$	$6\sqrt{F}$	$8\sqrt{F}$	$12\sqrt{F}$
Loop misclosure not to exceed (mm)	$4\sqrt{F}$	$5\sqrt{F}$	$6\sqrt{F}$	$8\sqrt{F}$	$12\sqrt{F}$

D, shortest length of leveling line (one-way) in km; E, shortest one-way length of section in km; F, length of loop in km.

The normalized residuals from a minimally constrained least-squares adjustment will be checked for blunders. The observation weights will be checked by inspecting the postadjustment estimate of the variance of unit weight. Elevation difference standard errors computed by error propagation in a correctly weighted least-squares adjustment will indicate the provisional accuracy classification. A survey variance factor ratio will be computed to check for systematic error. The least-squares adjustment will use models that account for:

Gravity effect or orthometric correction
Rod scale errors
Rod (Invar) temperature
Refraction-need latitude and longitude to 6″ or vertical temperature difference observations between 0.5 and 2.5 m above the ground
Earth tides and magnetic field
Collimation error crustal motion.

BIBLIOGRAPHY

FGCC. 1984. *Standards and Specications for Geodetic Control Networks.* Rockville, MD: Federal Geodetic Control Committee.
Hoskinson, A. and J. Duerksen. 1952. *Manual of Geodetic Astronomy: Determination of Longitude, Latitude, and Azimuth*, Special Publication 237. Washington, DC: U.S. Coast and Geodetic Survey, 205 pp.

Appendix C
Glossary of Terms

Aerial Survey: A survey or mapping project utilizing photographic, electronic, or other data obtained from an airborne sensor or platform. See also UAS Survey.

Airborne LIDAR: LIDAR is an acronym of "light detection and ranging" or "laser imaging, detection, and ranging." It is commonly used in airborne (aerial) surveys for making high-resolution maps but has other applications in ground-based measurements as well. For instance, it is used in augmented reality devices for onsite measurements of features, 3D laser scanning, altimetry, and other areas.

Augmented Reality: An augmented reality technology overlays digital information such as a virtual 3D model on a live image of what is being viewed through a device such as a smartphone camera. Primarily, an augmented reality system comprises a mobile device, handheld or head-mounted, with an integrated display screen, a camera, processor, a Global Positioning System (GPS) or GNSS, inertial orientation sensing, microphone, and other components.

Boundary Survey: A land survey to establish or reestablish a boundary line on the ground and/or to obtain data for a map or plot of land showing a boundary line.

Cadastral Survey: A land survey relating to land boundaries and subdivisions made to create land units suitable for transfer or to define the limitations of land title; derived from the word "cadastre" which has a dictionary meaning of "a register of property showing the extent, value, and ownership of land for taxation." In the United States, the term is used to refer to a survey which creates, marks, defines, retraces, or reestablishes the boundaries and subdivisions of the public land of the United States. See also Boundary Survey.

Construction Survey: Survey measurements made while construction is in progress, to control horizontal position and dimensions, elevation, and configuration; to determine adequacy of completion; and to obtain essential dimensions for computing construction quantities.

Control Survey: A survey project to provide horizontal and vertical position data for the support or control of subsequent surveys or mapping. The horizontal and vertical position data are provided in what is referred to as a datum (see datum—local, geodetic, elevation, mean sea level).

Continuously Operating Reference Station (CORS): A permanently installed/ constructed GNSS station with accurately known geodetic position, and continuously collecting carrier phase and code range measurements in support of 3D positioning activities. The data collected are transmitted and made available online or archived for later access by users. The data can be used as base station data in surveys where a user only needs one receiver to collect data on points and then post processes to obtain accurate position of the points in the same reference frame as the CORS stations. It is also used to broadcast corrections for real-time positioning applications without post-processing.

Datum: A survey datum is a horizontal or vertical reference system for making survey measurements and computations. A "horizontal geodetic datum" is defined using a mathematical figure such as an ellipsoid and can be realized by the latitude and longitude of one selected point in an area, and the azimuth from the selected point to an adjoining point. A vertical "geodetic datum" is defined relative to the mean sea level. "Elevation datum" is usually mean sea level as determined by hourly readings over an 18.6-year average. In small area projects, an assumed "local datum" (horizontal and/or vertical) is sometimes used.

Digital Elevation Model: A 3D digital representation of the variations in terrain elevation portrayed on a map or aerial imagery or aerial photography. It is also sometimes referred to as Digital Terrain Model.

Elevation: The orthometric height above a point of reference. For example, height above mean sea level, or vertical distance above or below the geoid.

Ellipsoid: A mathematical figure formed by revolving an ellipse about its minor axis. Two quantities define an ellipsoid, the length of the semi-major axis, a, and the flattening, f.

Engineering Survey: A survey project for the purpose of obtaining information essential for planning an engineering project or for developing and estimating its cost.

Geodetic Survey: A survey project which takes into account the figure and size of the Earth due to the large scale and extent of distances and areas involved in the measurements. It is the process of determining positions of points and/or distances (baselines) between points by making measurements to/from distant targets which must take into account the figure and size of the Earth. GNSS survey falls into this category because they use ranging from distant Earth-orbiting satellites.

GNSS Baseline: A baseline consists of a pair of stations for which simultaneous GNSS data have been observed and/or collected. It is mathematically expressed as a vector of coordinate differences between the two stations, or an expression of the coordinates of one station with respect to the other (whose coordinates are assumed known and is typically referred to as a "base" or "reference" station). See also Session.

GNSS Base Station: A base station, or reference station, is a GNSS receiver set up (permanently or temporarily) on a location with known position to collect data for differentially correcting data files of another receiver (which may be referred to as the "rover" receiver).

GNSS Survey: The process of determining positions of points on the Earth's surface by using artificial satellites, GNSS, as points to which, or from which, measurements are made.

Milestone, Project: Specific progress point in a project timeline. They mark progress needed to complete a project successfully.

Mining Survey: A survey project to determine the positions and dimensions of underground passages of a mine, including of the natural and artificial (surface and underground) features relating to the mine. The data collected during a mining survey include both horizontal and vertical positions, lengths, slopes and directions of tunnel, geologic and topographic characteristics of the vicinity, and ownership of the land and of the mine.

Online Positioning User Service (OPUS): A web-based GNSS data processing engine (software) by the U.S. National Geodetic Survey, which enables users to perform GNSS surveys without the need to, for example, occupy their own geodetic control or base stations for reference. The main advantage is that a user can simply occupy a project point with a single GNSS receiver, collect raw measurements, and then submit the RINEX data to OPUS to obtain accurately georeferenced positions. Multiple other functions are available through the web interface in support of geomatics projects. Other similar services exist worldwide, such as AUSPOS by Geoscience Australia and CSRS-PPP by the Canadian Geodetic Survey of Natural Resources Canada.

Project: A proposal, plan, or scheme of something to be done. A temporary, goal-driven effort to create a unique output or results, has clearly defined phases (milestones), and success is measured by whether it meets its stated objectives. An organized undertaking involving one or more tasks, usually with definite limits and specifications or standards, with start and end time. A special unit of work or research, usually involving many tasks, such as those in geomatics, surveying, engineering, construction, and aerial mapping.

Property Survey: A land survey project to determine boundary lines between privately owned parcels of land. See also Boundary Survey.

RINEX (Receiver INdependent EXchange): The GNSS raw data file format used for storing GNSS observations or measurements, containing time, carrier phase, and pseudorange observables. It was designed for free exchange of GNSS data and to be compatible with any, or most, processing software packages.

Session: In GPS/GNSS surveys, a project session refers to all the measurements taken simultaneously for a given duration. For example, a baseline session involves two receivers at two different stations (measurement points) receiving (and recording) signals from common satellites for the common duration. A network session is defined in a similar manner.

Specification: Specifications are the field operations or procedures required to meet a particular standard; the specified precision and allowable tolerances for data collection and/or application, the limitations of the geometric form of acceptable network figures, monumentation, and description of points.

Standard: An exact value, or concept thereof, established by authority, custom, or common consent, to serve as a rule or basis of comparison in measuring quantity, content, extent, value, quality, and capacity. In geomatics and land surveying, standards are typically defined as the minimum accuracies deemed necessary to meet specific objectives: a reasonably accepted error, a level of precision of closure, and a numerical limit on the uncertainty of coordinates.

Subdivision Survey: A type of land survey in which the legal boundaries of an area of land are located, and the area is divided into parcels of lots, streets, right-of-way, and other such things. All necessary corners or dividing lines (boundaries) are marked or monumented.

Task, Project: A unit of work or activity needed for progress toward project goals, and typically must be completed by a set deadline. Tasks may be further divided into subtasks or assignments.

Topographic Survey: A land survey project to determine the configuration (relief) of the surface of the Earth (land) and the location of natural and artificial objects.

Unmanned aerial system (UAS) Survey: A survey or mapping project utilizing photographic, electronic, or other data obtained from an airborne sensor on an UAS. It is also referred to by other terms such as Drone survey or unmanned aerial vehicle survey.

Universal Transverse Mercator (UTM): A global map projection coordinate system developed by the U.S. Military. It uses Transverse Mercator projection and divides the entire Earth into 60 zones. Each UTM zone is 60° wide and goes from pole to pole.

Index